LABORATORY BIORISK MANAGEMENT

Biosafety AND Biosecurity

LABORATORY BIORISK MANAGEMENT

Biosafety AND Biosecurity

Edited by

Reynolds M. Salerno

Sandia National Laboratories
Albuquerque, New Mexico, USA

Jennifer Gaudioso

Sandia National Laboratories
Albuquerque, New Mexico, USA

CRC Press
Taylor & Francis Group
Boca Raton London New York

CRC Press is an imprint of the
Taylor & Francis Group, an **informa** business

CRC Press
Taylor & Francis Group
6000 Broken Sound Parkway NW, Suite 300
Boca Raton, FL 33487-2742

First issued in paperback 2020

ISBN-13: 978-1-4665-9364-0 (hbk)
ISBN-13: 978-0-367-65882-3 (pbk)

Library of Congress Cataloging-in-Publication Data

Laboratory biorisk management : biosafety and biosecurity / editors, Reynolds M.
Salerno and Jennifer Gaudioso.
p. ; cm.
Includes bibliographical references and index.
ISBN 978-1-4665-9364-0 (hardcover : alk. paper)
I. Salerno, Reynolds M. (Reynolds Mathewson), 1966-, editor. II. Gaudioso,
Jennifer Marie, 1972-, editor.
[DNLM: 1. Laboratories. 2. Security Measures. 3. Bioterrorism--prevention &
control. 4. Toxins, Biological. QW 523]

R857.M3
610.28'4--dc23   2014047222

Visit the Taylor & Francis Web site at
http://www.taylorandfrancis.com

and the CRC Press Web site at
http://www.crcpress.com

Contents

Chapter 8   Evaluating Biorisk Management Performance 145

LouAnn Burnett and Patricia Olinger

Chapter 9   Communication for Biorisk Management 169

Monear Makvandi and Mika Shigematsu

Foreword

Communicable diseases remain a leading cause of death globally and account for nearly one-third of world deaths. The emergence of newly identified pathogens, as well as the re-emergence of pathogens with public health significance, exacerbates the global threat of infectious diseases. For example, it has been reported that between 1973 and 2003 over 36 newly emerging infectious diseases had been identified. Research and diagnostic activities involving pathogenic microorganisms are critical to global security as this research elucidates knowledge and leads to products that improve the health, welfare, economy, quality of life and security for all persons around the globe.

Advancements in technology as well as the cross-fertilization of formerly disparate scientific disciplines have led to technical capabilities never before realized in the life sciences. This technical progress is exemplified by the *de novo* synthesis of poliovirus and the recreation of the 1918 H1N1 influenza virus, which was the causative agent of the *Spanish flu* pandemic, the deadliest single event in recorded history killing an estimated 50 million people world-wide.

In addition to the threat to public health and welfare caused by pathogenic microorganisms derived from nature, including newly emerging or re-emerging diseases, there is also the threat posed by the intentional release of disease-causing microorganisms whether through state-sponsored biological warfare or through the intentional use of pathogens to elicit terror. The impact of such a terrorist event was demonstrated vividly in 2001 during the Amerithrax episode. Highly refined (i.e., weaponized) spores of *Bacillus anthracis* were released on an unsuspecting public, resulting in five deaths, illness in 17 U.S. citizens, and an untold economic impact.

The combined threats to public health resulting from emerging diseases and the potential for deliberate release of a pathogenic microorganism altered the research and public health agenda not only for the U.S., but also for countries around the globe. For example, in 2003 the U.S. National Institute of Allergy and Infectious Diseases (NIAID) established Regional Centers of Excellence for Biodefense and Emerging Infectious Diseases Research to serve as regional foci for developing and conducting cutting edge research. The centers were created to develop countermeasures to these threats, including vaccines, therapeutics, and diagnostics, among others. At the same time, the U.S. Department of Homeland Security provided financial support to fund the construction and expansion of a laboratory infrastructure to support this infectious diseases research agenda. This expansion of infrastructure and funding was not unique to the U.S., and can be observed internationally.

Concerns raised by the threat of potential biological terrorism in the national security apparatus of the U.S. and other countries resulted in the promulgation of regulations intended to control and limit the numbers of persons with access to certain pathogenic microorganisms. This regulatory approach focused on the establishment of a security-based infrastructure and security-based programs to manage the important research directed toward understanding the fundamental biology of, and generating medical countermeasures against, certain dangerous pathogens.

Biosafety and biosecurity, while distinctly different concepts, are inexorably linked; one cannot consider a biosafety program to be robust in the absence of bio-security, and most certainly, biosecurity cannot exist in the absence of a strong com-mitment to biosafety. It has become abundantly clear that a holistic approach to the management of risks associated with research involving pathogenic microorgan-isms is critical. Facilities and infrastructure construction and maintenance, educa-tion, training and competency (not only of the scientific staff but also all support staff ancillary to the research program), reliability of the entire workforce, public outreach and political support, strong leadership committed to the management of biorisk (i.e., biosafety and biosecurity), and a culture of responsible research are all elements that must be integral to life sciences research, especially this particular research enterprise.

In the early to mid-1970s, a new technology, termed "recombinant DNA technology," was developed and utilized. The scientific community responsible for developing and utilizing this work also realized that the new technology posed potential risks and threats to the health and well being of society. To address the pub-lic and political concerns, the scientific community in the U.S. came together to craft guidelines within the structure of the National Institutes of Health (*Guidelines for Research Involving Recombinant DNA Molecules*) by which biosafety risks from this research could be systematically assessed and through which these biosafety risks could be specifically mitigated and managed. These guidelines were not prescrip-tive, but rather were performance based, allowing for flexibility in the manner by which these risks could be mitigated. Most critically, the guidelines provided mech-anisms for local oversight by the scientists themselves, their research institutions, and the funding agency. While this approach toward management of biosafety risks associated with research involving pathogenic microorganisms is clearly important, this approach alone is incomplete as it does not address biorisk in a holistic manner.

This book proposes a new paradigm for evaluating, mitigating, and managing biorisk and terms this paradigm *AMP: Assessment/Mitigation/Performance*. While specific individual components of this new paradigm are currently being employed in biosafety programs built upon the existing *"biosafety level"* systems, many bio-safety programs fail to comprehensively approach risk assessment, risk mitigation, and performance evaluation. For example, a comprehensive risk analysis of both bio-safety and biosecurity (*biorisk*) is generally lacking in these traditional approaches. Similarly, where traditional biosafety level-based systems discuss levels of controls to mitigate risk (usually built upon *mitigation control measures* that include engi-neering controls, administrative controls, practices and procedures, and personal protective equipment), few *routinely evaluate* the effectiveness of these risk mitiga-tion strategies.

In the Summer of 2014 several highly publicized incidents and accidents involving the potential release of some of the world's most dangerous pathogens (e.g., smallpox, *Bacillus anthracis,* Highly Pathogenic Avian Influenza H5N1, and Ebola) from the laboratories of several U.S. Federal agencies resulted in a strong negative response from the public, who were understandably fearful of the threat to public health posed by these releases. Following on fears of the citizenry and resultant backlash against the scientific community conducting this research, political pressure on these same government agencies resulted in a funding pause in the U.S. of important research involving influenza as well as SARS and MERS coronaviruses. Investigations into the root causes leading to these accidents and incidents revealed that a contributing factor may have been the prioritization of security procedures over safety practices.

As a result of these incidents, the Secretary of the U.S. Department of Health and Human Services ordered an external review of safety programs in DHHS labs, including the CDC. A report to the CDC by this independent external advisory group conveyed many of the same observations made about these incidents and detailed in Chapter 10 of this book. The case study analyses of these incidents provided in this chapter concluded that a comprehensive approach to biorisk management was absent at the CDC at the time of these incidents. In fact, many of the recommendations proposed by the external advisory group are consistent with and reflect the comprehensive approach to risk management presented in this book.

As already stated, research activities involving pathogenic microorganisms elucidate basic knowledge and lead to products and technologies that improve the welfare, economy, and quality of life for people globally. It is important that the scientific community embrace a holistic approach to the management of biorisks because biorisk management is a responsibility shared by principal investigators, bench scientists, support staff, students, postdoctoral fellows, and the leadership of the institutions conducting and funding this vital research. Furthermore, it is equally important that the scientific community speak loudly and proudly of the benefits of their research activities to educate and gain public support for, and acceptance of, this work. It is also vital to inspire youth to become conversant in and enthusiastic about the benefits of basic science.

Joseph Kanabrocki, Ph.D., SM(NRCM)
Associate Vice-President for Research Safety
Professor of Microbiology
University of Chicago

Preface

The central premise of this book is that the biological research, clinical, diagnostic, and production/manufacturing communities need to embrace and implement biorisk management systems in their facilities and operations. In most countries, the current system mitigates the risk of accidental infection, accidental release, and intentional misuse of pathogens and toxins based on general, predetermined biosafety levels and/or prescriptive biosecurity regulations. Although this approach may have sufficed when the biological life science community was relatively small, and work with particularly dangerous organisms was limited to a few countries and facilities, the life sciences have grown significantly in the last two decades—in both scope and sophistication. Much of this growth has extended well beyond North America and Western Europe, and deep into the developing world. Simply, there are more people in more places working with, and even creating, more dangerous pathogens and toxins than ever before—and that trend shows no sign of abating in the future.

Increased risk inevitably comes with this expansion. The past 20 years have been marked by multiple safety and security incidents at bioscience facilities around the world, including many notable incidents at so-called sophisticated facilities in North America and Western Europe. Clearly, the current system that is based on biosafety levels and security regulations does not work effectively enough. It is time for the bioscience community to learn some lessons from other high-consequence industries that have experienced devastating accidents, and that have intellectually evolved their own approaches to safety and security well beyond generic, predetermined, technical checklists. These industries have almost universally adopted what is now referred to as risk management systems. This book urges the global bioscience community to embrace biorisk management—before a devastating accident threatens to jeopardize the entire bioscience enterprise.

This book is organized into 11 separate chapters, and each chapter focuses on a different element of a biorisk management system. Different experts from around the world have written each chapter, demonstrating that the biorisk management system espoused by this book is globally applicable. The first chapter defines biorisk management, details the history of the field of biosafety and biosecurity, and makes a case for implementing biorisk management to prevent a major incident by drawing comparisons to disasters in other industries. The second chapter describes the AMP model, which is a framework that uses the components of assessment, mitigation, and performance to structure and implement a comprehensive biorisk management system. Chapter 3 defines the risk assessment process, and explains how to assess and prioritize various risks, and ensure that a risk assessment fits into a biorisk management system. The fourth chapter illustrates how to use the risk assessments to inform a design strategy to avoid overengineering a facility and wasting valuable resources.

The fifth chapter evaluates the roles of the different mitigation measures, including laboratory practices and procedures, safety and security equipment, and personnel management. The specific combination of mitigation measures should be determined based on the risk assessment, and evaluated according to specific performance metrics. Chapter 6 argues that a flexible and adaptable training plan is more effective than a rigid standardized compliance plan, since it can be strategically implemented to manage risk in a number of different settings and contexts with the ability to meet the challenge of new hazards or threats.

Chapter 7 argues that reliability-centered maintenance should be the framework for a biorisk management maintenance program. The eighth chapter advocates for utilizing specific performance indicators, instead of relying on failure data, for proactive activities and outcomes to make effective changes and improvements to a biorisk management system. The ninth chapter suggests that a comprehensive biorisk management system must include a risk communication plan designed to address both normal operations and emergency situations from an internal and an external perspective.

Chapter 10 is a case study that examines the biosafety incidents that took place at the US Centers for Disease Control and Prevention and the US National Institutes of Health in 2014. Finally, Chapter 11 identifies some of the most important challenges that face the biorisk management community by examining current gaps and shortcomings in contemporary biorisk management understanding and approaches. It also presents a series of opportunities to enhance the practice of biorisk management in the future.

Acknowledgments

This book could not have been written or even attempted without the tremendous hard work, commitment, and support of so many of our mentors, colleagues, friends, and families. Our work around the world for more than the last decade has shaped our evolving views of biorisk management. We have not created the field of biorisk management, but have been fortunate to be among those who have helped shape its current form. Nevertheless, there are hundreds of experts around the world who successfully manage the risks of working with dangerous biological materials on a daily basis, and we have had the privilege to learn from the experiences of many of them.

In particular, we are grateful for the support and kindness offered to us by so many technical experts in our field, who have taught, guided, and encouraged us to contribute to the intellectual debate of how best to manage the risks of working with biological materials in research, diagnostic, and clinical laboratories, as well as in healthcare and field settings. Many of those experts have contributed chapters to this book. Among those who are not authors here, but deserve our recognition and deep appreciation, are Stefan Waggener, Paul Huntly, Vips Halkaer-Knudsen, Jim Welch, Maureen Ellis, Bob Ellis, Nicoletta Previsani, Ingegerd Kallings, Joe Kozlovak, Debra Hunt, and Heather Sheeley. We have been significantly influenced by all of these mentors, and we are humbled to be able to call them colleagues.

We are also grateful for the collaborations that we have had with the American Biological Safety Association, the European Biosafety Association, the Asia-Pacific Biosafety Association, the International Federation of Biosafety Associations, the World Health Organization, the World Organization for Animal Health, and the European Committee for Standardization. The US Department of State's Biosecurity Engagement Program and the US Defense Threat Reduction Agency's Cooperative Biological Engagement Program have given both of us, as well as our program at Sandia, the opportunity to work with hundreds of institutions and thousands of scientists from around the world to advance biorisk management. All of these international engagements on behalf of the US government have also helped shape our views on this topic.

This book is the product of countless discussions with the staff who work for Sandia's International Biological Threat Reduction (IBTR) program. Several IBTR staff contributed to these chapters as authors, and several other IBTR staff extensively peer-reviewed drafts of the chapters. The IBTR team is too large to mention every one of them here, but we are grateful for their tireless and global commitment to the pursuit of biorisk management. A few in particular, though, deserve special mention. Jason Bolles, Lyle Beck, and Laurie Wallis helped create the figures and table for this book. Laurie Wallis single-handedly managed the entire book project from beginning to end, cajoling the authors to complete their chapters, keeping the editors on task, negotiating with lawyers, ensuring proper reviews and approvals, and corresponding with the staff at CRC Press. Without Laurie, this book would not have seen the light of day; we cannot express enough gratitude for Laurie's efforts.

Sandia National Laboratories provided the financial support for this book. In particular, we express our appreciation to Vice President Jill Hruby and Director Rodney Wilson for supporting and funding this project. We also thank Greg Doudnikoff, an intellectual property attorney at Sandia, who executed the contract with CRC Press. We are particularly grateful to Rodney Wilson, director of Sandia's Center for Global Security and Cooperation, for his critical technical guidance and enduring programmatic support for our work in biorisk management.

We also received fantastic assistance and encouragement from our editors at CRC Press. Executive Editor Barbara Norwitz and Project Coordinator Kat Everett have been wonderful to work with throughout the duration of this project. Their patience, forbearance, encouragement, and expertise have been invaluable.

This book would not have been possible without the very significant contributions of all of these great mentors and colleagues, and we cannot overstate our deep gratitude to all of them. Nevertheless, all of the mistakes and shortcomings of this manuscript remain ours and ours alone.

Finally, we extend our deepest gratitude to our significant others, Jennifer Salerno and Damian Donckels. Each of them has sacrificed by allowing us to dedicate our professional lives to advancing biorisk management. Without their support and encouragement, this book would not have come to fruition.

Albuquerque, New Mexico

About the Editors

Reynolds M. Salerno is the senior manager for Cooperative Threat Reduction Programs at Sandia National Laboratories in Albuquerque, New Mexico. His programs enhance US and international security by reducing biological, chemical, and nuclear threats worldwide. Recognized as a leading expert on laboratory biosecurity, Salerno and his Sandia team have worked extensively in laboratory biosafety, biosecurity, biocontainment, and infectious disease diagnostics and control internationally. Salerno is a coauthor of the *Laboratory Biosecurity Handbook* (CRC Press, 2007). As a technical advisor to the World Health Organization (WHO), he

was a member of the WHO's international team that inspected the Maximum Containment (smallpox) Laboratory at the State Research Center of Virology and Biotechnology VECTOR, Koltsovo, Novosibirsk, Russia, in December 2009. He is a principal developer of the WHO train-the-trainers course, "Biorisk Management Advanced Training Program," which was delivered in all six WHO regions in 2010. He was recently the vice chairman of the board of directors of the International Federation of Biosafety Associations. Salerno received his PhD from Yale University in 1997.

Jennifer Gaudioso leads the International Biological Threat Reduction (IBTR) and International Chemical Threat Reduction (ICTR) programs at Sandia National Laboratories in Albuquerque, New Mexico. These programs enhance US and international security by promoting safe, secure, and responsible use of dangerous biological and chemical agents. They have organized many international conferences, trainings, and workshops to build local capacity to address these issues. The team currently consults in more than 40 countries specifically on biosecurity and chemical security issues.

Gaudioso and her Sandia team work with international partners, such as the World Health Organization and the International Federation of Biosafety Associations. Her program is a World Organisation for Animal Health (OIE) Collaborating Centre for Laboratory Biorisk Management. Gaudioso has served on the National

Academies' Committee on Education on Dual Use Issues in the Life Sciences and their Committee on Anticipating Biosecurity Challenges of the Global Expansion of High Containment Biological Laboratories. She is the author of numerous journal articles and has presented her research at national and international meetings. She also coauthored the *Laboratory Biosecurity Handbook* (CRC Press, 2007). Gaudioso has served on Sandia's Institutional Biosafety Committee, and is an active member of the American Biological Safety Association. She earned her PhD in chemistry at Cornell University.

Contributors

William D. Arndt
Sandia National Laboratories
Albuquerque, New Mexico

Lisa Astuto Gribble
Sandia National Laboratories
Albuquerque, New Mexico

Susan Boggs
Sandia National Laboratories
Albuquerque, New Mexico

Stefan Breitenbaumer
Spiez Laboratory
Spiez, Switzerland

Benjamin Brodsky
Sandia National Laboratories
Albuquerque, New Mexico

LouAnn Burnett
Sandia National Laboratories
Albuquerque, New Mexico

Susan Caskey
Sandia National Laboratories
Albuquerque, New Mexico

Ross Ferries
HDR Architecture, Inc.
Atlanta, Georgia

Mark E. Fitzgerald
HDR Architecture, Inc.
Atlanta, Georiga

Jennifer Gaudioso
Sandia National Laboratories
Albuquerque, New Mexico

Lora Grainger
Sandia National Laboratories
Albuquerque, New Mexico

Natasha K. Griffith
University of California–Los Angeles
Los Angeles, California

Hazem Haddad
Jordan University of Science and
   Technology
Irbid, Jordan

Laura Jones
Sandia National Laboratories
Albuquerque, New Mexico

Ephy Khaemba
International Livestock Research
   Institute
Nairobi, Kenya

Daniel Kümin
Spiez Laboratory
Spiez, Switzerland

Monear Makvandi
Sandia National Laboratories
Albuquerque, New Mexico

Sergio Miguel
Medical Forensic Laboratory
Buenos Aires, Argentina

Uwe Müeller-Doblies
Epibiosafe
Surrey, United Kingdom

Patricia Olinger
Emory University
Atlanta, Georgia

William Pinard
Sandia National Laboratories
Albuquerque, New Mexico

Reynolds M. Salerno
Sandia National Laboratories
Albuquerque, New Mexico

Edgar E. Sevilla-Reyes
National Institute of Respiratory
  Diseases
Mexico City, Mexico

Mika Shigematsu
National Institute of Infectious
  Diseases
Tokyo, Japan

Edith Sangalang Tria
San Lazaro Hospital
Ministry of Health
Manila, Philippines

Dinara Turegeldiyeva
Kazakh Science Center for Quarantine
  and Zoonotic Diseases
Almaty, Kazakhstan

Laurie Wallis
Sandia National Laboratories
Albuquerque, New Mexico

Cecelia V. Williams
Sandia National Laboratories
Albuquerque, New Mexico

1 Introduction
The Case for Biorisk Management

Reynolds M. Salerno and Jennifer Gaudioso

ABSTRACT

This introductory chapter recounts the history of laboratory biosafety and biosecurity in an attempt to explain the origins of the current paradigm that relies on pre-defined biosafety levels, agent risk groups, and biosecurity regulations. This history reveals that the fundamental concepts of biorisk management were articulated well before the current paradigm came into existence, but unfortunately have been lost by a community that has expanded significantly. After summarizing many safety and security incidents at bioscience laboratories in the 1990s and 2000s, demonstrating the weakness of the current paradigm, this chapter argues that the rapid technological advances of the biosciences compel the community to reconsider the traditional methods of ensuring safety and security. The chapter then reviews a series of catastrophic accidents that occurred in many different industries, and shows, as a result, how generic, rule-based, administrative systems have been abandoned in favor of a performance-based, holistic, risk-management systems approach. The implementation of substantive risk management policies, standards, and expectations has dramatically decreased both the number and severity of accidents in these industries. The bioscience community should not wait for a calamity to occur in its field before learning these fundamental lessons.

LABORATORY BIOSAFETY AND BIOSECURITY

Biorisk management encompasses both laboratory biosafety and biosecurity. The laboratory biosafety community has relied on predefined biosafety levels[*] for more than 30 years. Laboratory biosecurity has a much shorter history than biosafety, but it has been predominantly based on prescriptive regulations. In both cases, biosafety and biosecurity practices have generally relied on generic biological agent risk

---

[*] According to the WHO *Laboratory Biosafety Manual*, 3rd edition (2004), "Laboratory facilities are designated as basic—Biosafety Level 1, basic—Biosafety Level 2, containment—Biosafety Level 3, and maximum containment—Biosafety Level 4. Biosafety level designations are based on a composite of the design features, construction, containment facilities, equipment, practices, and operational procedures for working with agents from the various risk groups."

1

groups, biosafety levels, or regulations that, de facto, assume that all work with the same agent presents the same degree of risk—regardless of the nature of the work, where it takes place, or by whom. Diagnostic work with avian influenza should take place in essentially the same laboratory in Minneapolis as in Jakarta, and if the laboratories physically look the same, the risk of accidental release will be essentially the same. Of course, anyone with any substantive experience in the biological research and diagnostics field, particularly biosafety experts, will recognize the absurd simplicity of that statement, but for many who want to build a new bioscience facility, the published guidance leads them to believe that achieving the prescribed biosafety level equates to biological safety in that facility.

The situation is arguably worse for laboratory biosecurity. Because most professionals in the bioscience community have little experience or expertise in law enforcement or counterterrorism, policy makers with scant bioscience background have enacted regulations that define the technical security systems for every facility that works with specific agents or toxins. The unique circumstances of the facility or its personnel, its location, the agents, and the nature of the work with those agents seem not to matter. From the perspective of the regulators, all the facilities that work with certain dangerous agents in their jurisdiction should employ the same security approach. Not only does this simplified approach inevitably lead to wasted security resources in some facilities, and significant gaps in security at other facilities, but most disturbingly, it discourages the scientific leadership to engage intellectually on the subject of security. Such an attitude inevitably leads to complacency and increases a facility's vulnerabilities over time.

A Short History of Laboratory Biosafety

It is instructive to understand the history of laboratory biosafety and biosecurity. Much of the published history is rooted in North America and Western Europe. Biosafety as an intellectual field can trace at least many of its origins to the US biological weapons program, which was active during the Cold War and finally terminated by President Nixon in 1969. In 1943, Ira L. Baldwin became the first scientific director of Camp Detrick (which eventually became Fort Detrick), and was tasked with establishing the biological weapons program (US Department of Army 2014). The US development of biological weapons was explicitly for defensive purposes: to enable the United States to respond in kind if attacked by such weapons. After the Second World War ended, Camp Detrick was designated a permanent installation for biological research and development. Baldwin understood from the very beginning that the program had to establish specific measures to protect Camp Detrick personnel and the surrounding community from the dangers of the highly infectious agents that the program would work with on a daily basis. Biosafety was necessarily an inherent component of biological weapons development.* Baldwin immediately assigned Newell A. Johnson to design any needed modifications for safety. Johnson engaged some of Camp Detrick's leading scientists about the nature of their

---

* It is fair to assume that, at this time or even before, the British, Soviets, and Japanese also had developed biosafety measures as part of their biological weapons development programs.

work, and developed specific technical solutions—such as Class III safety cabinets and laminar flow hoods—to address their specific risks (US Department of Army 2014). Over time, Johnson and his colleagues recognized the need to share their technical challenges and solutions with other facilities that were also part of the US biological weapons program, and in 1955 they began to meet annually to discuss biosafety issues. This annual meeting eventually led to the formation of the American Biological Safety Association (ABSA) in 1984, and the annual meeting soon became the ABSA annual conference (Barbeito and Kruse 2014).

Although the US biological weapons offensive and defensive programs are among the most documented contributions to a systematic approach to developing biosafety, these pioneers recognized the contributions of others to the field. For example, Arnold Wedum cites descriptions of the use of mechanical pipettors to prevent laboratory-acquired infections in German scientific journals that date back to 1907 and 1908 (Wedum 1997). Ventilated cabinets, early progenitors to the nearly ubiquitous engineered control now known as the biological safety cabinet, were also first documented outside of the US biological weapons program. A pharmaceutical company in Pennsylvania developed a ventilated cabinet for work with *Mycobacterium tuberculosis* in 1909 (Kruse et al. 1991). In 1954, tuberculosis was the driving factor that led to the adoption of similar ventilated cabinets at the Goteborg Bacteriological Laboratory (Lind 1957). These early efforts helped the bioscience community begin to more broadly adopt principles of this nascent field of biosafety.

At approximately the same time as the United States formally abandoned its biological weapons program, the international community aggressively pursued the eradication of smallpox (College of Physicians of Philadelphia 2014).* The smallpox eradication campaign, which officially began in 1967, also had a significant impact on the evolution of the biosafety field. Between 1963 and 1978, there were a series of smallpox infections that originated from laboratories in the United Kingdom. In that time period, the United Kingdom had 80 cases of smallpox infections that were traced to two accredited smallpox laboratories (Shooter 1980; Furmanski 2014). The most egregious of these incidents occurred in August 1978—well after the last wild case of Variola major in 1975, and the last wild case of Variola minor in 1977. Janet Parker, a medical photographer at the University of Birmingham Medical School, worked in a darkroom one floor above a laboratory where research was being conducted with live smallpox virus. After contracting the disease at work, and then infecting her mother, Parker became the last person known to die from smallpox. Her mother survived, and 300 of her colleagues and contacts were quarantined. Prior to this event, the World Health Organization (WHO) had informed Henry Bedson, the head of the medical microbiology department, that his facilities did not conform to the WHO guidelines. Bedson failed to make any of the WHO's recommended changes in lab procedures. Shortly after Parker's death, Bedson committed suicide, purportedly over the guilt for his part in the tragedy (College of Physicians of Philadelphia 2014).

---

* The World Health Organization launched the Intensified Smallpox Eradication Programme in 1967. At the time, smallpox was endemic in 12 countries or territories in eastern and southern Africa, 11 in Western and Central Africa, seven in Asia, and Brazil in the Americas. The World Health Assembly declared smallpox eradicated in 1980.

The accidental infection and death of a laboratory worker and the secondary transmission to someone outside of the laboratory—after the international community had spent US$23 million per year between 1967 and 1979 to eradicate smallpox (Center for Global Development 2014)—raised serious concerns about biosafety practices worldwide (e.g., Pike 1976*), and contributed directly to the decision by the World Health Assembly to consolidate the remaining stocks of smallpox into two locations: the Centers for Disease Control and Prevention (CDC) in the United States and the State Research Center of Virology and Biotechnology (known as VECTOR) in Russia. In addition, this tragedy spurred forward some embryonic biosafety initiatives. In 1974, the CDC had published the *Classification of Etiological Agents on the Basis of Hazard* (US Centers for Disease Control and Prevention 1974), which introduced the concept of establishing ascending levels of containment that correspond to risks associated with handling groups of infectious microorganisms that present similar characteristics—or so-called agent risk groups. Two years later, the US National Institutes of Health published the NIH Guidelines for Research Involving Recombinant DNA Molecules (US National Institutes of Health 1976), which described in detail the microbiological practices, equipment, and facility safeguards that correspond to four ascending levels of physical containment.

These guidelines set the foundation for a code of biosafety practice that was formally introduced in 1983, when the WHO published the first edition of its *Laboratory Biosafety Manual* (World Health Organization 1983), and in 1984, when the CDC and NIH jointly published the first edition of *Biosafety in Microbiological and Biomedical Laboratories* (BMBL) (US Department of Health and Human Services 1984). These documents established the paradigm of biosafety containment levels that should be implemented for work with certain agents. Increasing biosafety levels were designated for biological agents that pose increasing risk to human health. Understandably, the biosafety levels focused on the technical means of mitigating the risk of accidental infection or release. Although the biosafety levels articulated a combination of engineered controls, administrative controls, and practices, the emphasis was clearly on equipment and facility controls. Specific technologies and physical barriers were associated with each of the biosafety levels, and risk assessment was not emphasized. In fact, the implication was that the risk assessment had already been done by the experts, who had categorized the agents into risk groups, and established specific controls for each of the biosafety levels. A community of "biosafety officers" soon emerged, who adopted the administrative role in bioscience facilities of ensuring that the proper equipment and facility controls were in place based on the specified biosafety level of the laboratory.

Unfortunately, the biosafety levels paradigm that was adopted in the early 1980s seemed to have overlooked some seminal work on biosafety that first appeared more than two decades earlier. In 1961, Brooks Phillips published a study of biosafety based on his visits to 102 laboratories in 11 countries, and concluded that preventing accidental releases and infections in laboratories required a broad, systematic approach that should not be limited to the implementation of technical control measures:

---

* This seminal study documented 3,921 historical cases, of which 2,465 occurred in the United States and 164 were fatal.

A 'whole laboratory' concept can be evolved in which importance may be attached to such varied sub-components as management, training, building construction, air ventilation and filtration, disinfectants, immunization, and the use of special equipment and techniques…. One may ask a number of pertinent questions. To what extent is microbiological safety needed in various types of laboratories? Which sub-components are most important? Is the application of these newer developments fully justified from the point of view of costs? (Phillips 1961)

Furthermore, Phillips concluded that management displayed evidence of its safety responsibilities in only 43% of the facilities that he visited. Without explicit management direction and support, the implementation of any biosafety controls would not be reliably or consistently effective. And conversely, leadership on discrete safety issues had a significant positive influence on staff choosing to use available risk mitigation measures. For instance, the risk of mouth pipetting infectious solutions was nearly universally recognized. But in laboratories whose leadership did not insist on the use of mechanical pipettors, staff defaulted to mouth pipetting whenever they felt too busy to use the "more time-consuming" mechanical pipettors. When the lab leadership enforced the requirement to use mechanical pipettors instead of mouth pipetting, there was good compliance, few complaints, and technicians were proud that the director was concerned about their health (Phillips 1961).

Phillips also illustrated how the laboratory design impacts safety, citing building features such as size and shape, room size and layout, ventilation, and separation of infectious areas. He evaluated a series of laboratory designs based on his visits to demonstrate the effect of each on biosafety, and he argued that many of the most recently constructed laboratories did not optimize their design or operations with biosafety as a priority. He articulated challenges that, unfortunately, remain familiar to today's lab directors: "They frequently find themselves without sufficient information on laboratory hazards, on the frequency of laboratory illness, or about recent developments in building design—information which is needed to present cogent arguments for increased building funds" (Phillips 1961). As far back as 1961, Phillips recognized that a well-designed building can and should support the safe and secure execution of work by the end user. The building layout, and resulting workflows, is the defining element in the creation of public and private zones that are critical for both safety and security.

Arnold Wedum, director of Industrial Health and Safety at the US Army Biological Research Laboratories from 1944 to 1969, has been recognized as one of the pioneers of biosafety who, according to the BMBL, "provided the foundation for evaluating the risks of handling infectious microorganisms and for recognizing biological hazards and developing practices, equipment, and facility safeguards for their control" (US Department of Health and Human Services 2009). In 1966, Wedum and his Fort Detrick colleague, microbiologist Morton Reitman, analyzed multiple epidemiological studies of laboratory-based outbreaks, and showed that no infections occurred in people outside of the building that were not associated with the laboratory. They concluded that primary containment is sufficient for most research and diagnostic activities, and filtering the exhaust air is unnecessary except for situations such as laboratories that work with dry micronized microbial particles or pilot plants that grow pathogens in aerated tanks with agitators. The fundamental point articulated by Wedum and Reitman was that laboratories should be designed based on a risk assessment that is specific for the work to be conducted at that particular facility (Reitman and Wedum 1966).

Wedum and Reitman were not alone in promoting this perspective. As far back as 1954, Rolf Saxholm identified the risk of laboratory-acquired infection of tuberculosis during centrifugation. To mitigate this risk, he developed a procedure that eliminated the need to centrifuge tuberculosis specimens. Although he demonstrated the utility of his method, his procedure never became widely accepted outside of Norway (Saxholm 1954). In 1961, Brooks Phillips documented how different laboratories developed different technical solutions to effectively mitigate the same risks. For example, to mitigate the risk of aerosol exposure to tuberculosis during centrifugation of sputum samples, the Imperial Chemical Industries Laboratory in the United Kingdom located their centrifuges in ventilated hoods, while other laboratories placed centrifuge tubes in protective cases before centrifuging (Phillips 1961).

Despite the publication of these foundational biosafety works, the predefined, biosafety-levels paradigm that was promulgated almost 20 years later seemingly eliminated the expectation for a site-specific and work-specific risk assessment. Instead, a facility operator could rely on the predetermined agent risk groups and biosafety levels for design guidance. Risk assessments became the equivalent of identifying the material safety data sheet of the agent that would be used in a particular experiment. Comprehensive risk assessments, as advocated by Wedum and Reitman, became increasingly rare. And, as a result, many laboratories in the United States and elsewhere were "overdesigned," wasting precious resources that could have been allocated to other, perhaps more effective, biosafety control measures.

A Short History of Laboratory Biosecurity

This agent- or prescribed level-based risk perspective bled over into the laboratory biosecurity realm once the concern about the misuse of dangerous pathogens became more commonplace. After Larry Wayne Harris ordered *Yersinia pestis* under false pretenses, the US government enacted the so-called select agent regulations in 1996 to regulate the transfer of a select list of biological agents from one facility to another (US Code of Federal Regulations 1996). If a facility transferred an agent on this list, then the regulations determined that there was risk of misuse. If a facility did not ship a pathogen on the government list, then the government believed there was no risk of misuse. The US government changed its perspective slightly after the terrorist attacks and the Amerithrax attacks of 2001: the revised select agent regulations now required specific security measures for any facility in the United States that used or stored one or more agents on the new, longer list of agents (US Code of Federal Regulations 2005). Again, however, the government took responsibility for determining the risk, and the risk was deemed the same for all the agents on the list. The security risk of an agent not on the list was determined to be so low or nonexistent that security measures were not required for that particular agent. After years of complaints from the bioscience community about this simplistic dichotomy between biosecurity risk for those agents on the list and no biosecurity risk for those agents off the list, the select agent regulations were revised in 2012 to create two tiers of select agents: Tier 1 agents that pose the greatest risk of deliberate misuse, and the remaining select agents. This change was intended to make the regulations more risk based, mandating additional security measures for Tier 1 agents (US White House 2010). Yet the explicit message to the bioscience community remained that it was not necessary for individual facilities to conduct their own security risk assessments, or to design the most appropriate security systems to mitigate their unique risks. Instead, the message was that compliance with regulations was a sufficient form of risk mitigation. Security risk management was hardly necessary.

Other countries have also implemented relatively simplistic and prescriptive biosecurity regulations for bioscience facilities. Singapore's Biological Agents and Toxins Act is similar in scope to the US regulations, but with more severe penalties for noncompliance (Republic of Singapore 2005). South Korea amended its Act on Prevention of Infectious Diseases in 2005 to require institutions that work with listed "highly dangerous pathogens" to implement laboratory biosafety and biosecurity requirements to prevent loss, theft, diversion, release, or other misuse (Government of South Korea 2005). Under Japan's recently amended infectious disease control law, Japan's Ministry of Health, Labor, and Welfare established four schedules of select agents that are subject to different reporting and handling requirements for possession, transport, and other activities (Government of Japan 2007). Canada certifies Canadian containment level (CL) 3 and CL4 facilities that work with risk group 3 or 4 human pathogens (Public Health Agency of Canada 2014). In 2008, the Danish Parliament passed a law that gives the Minister of Health and Prevention the authority to regulate the possession, manufacture, use, storage, sale, purchase or other transfer, distribution, transport, and disposal of listed biological agents (Kingdom of Denmark 2008).

But in all cases, the paradigm of regulated security measures based on a pre-defined list of agents is the same. All a facility must do, if it is subject to these bio-security regulations, is implement prescribed security measures if it works with or stores an agent on the list. Biosecurity implementation has become a purely administrative activity based on a government-developed checklist. Risk assessment and management by the principals who work with, own, and are ultimately responsible for those agents are superfluous activities. Why do it if it is not required?

RECENT BIOSAFETY AND BIOSECURITY INCIDENTS

We believe that the bioscience community depends too heavily on predefined solutions sets, known as agent risk groups, biosafety levels, and biosecurity regulations. This dependence has relegated laboratory biosafety and biosecurity to the administrative basements of bioscience facilities. These generic agent risk groups, biosafety levels, and biosecurity regulations have almost eliminated the pursuit of the intellectually rigorous, risk-based assessments and solutions of the 1960s—when the field was in its infancy. Instead, we now often have complacency in laboratory biosafety and biosecurity, and the general absence of comprehensive management systems to mitigate these risks.

Such complacency has contributed directly to a series of safety and security incidents in bioscience facilities, especially since the field began to expand and advance significantly in the mid to late 1990s. The nature of these incidents demonstrates the fundamental weakness of the biosafety and biosecurity paradigm based on agent risk groups, biosafety levels, and biosecurity regulations. The increasing frequency of these incidents portends disaster for the bioscience field unless the biosafety and biosecurity paradigm changes dramatically.

There have been many recent incidents of laboratory-acquired infections that can be attributed to a failure to wear basic personal protective equipment and follow simple good laboratory practices. For instance, in 2001, the *New England Journal of Medicine* published a report of the first human case of glanders in the United States in over 50 years. A 33-year-old microbiologist, who worked with *Burkholderia mallei* at the US Army Medical Research Institute of Infectious Diseases (USAMRIID) in Frederick, Maryland, did not routinely wear gloves. It is believed that his exposed skin was the means of infection. His illness persisted and grew more severe after several months, and his treatment was complicated by the lack of clinical experience with glanders (Srinivansan et al. 2001). In 1996, 6 of 19 medical technologists who worked in a clinical microbiology laboratory in Rhode Island were infected and became ill with *Shigella sonnei*. Study of the cultured isolates indicated that the *Shigella* strain in question was nearly identical to a control strain kept by the laboratory, and which was in use at the time of exposure by one of the unaffected medical technology students. The student was the only member of the laboratory to routinely wear gloves. However, he did not follow other laboratory protocols, including the use of a separate processing sink for disposal of work samples. Instead, he utilized a more convenient hand washing sink, which he contaminated with *S. sonnei*. In turn, his colleagues who used the sink's faucet handles were infected with *S. sonnei*. If laboratory management had insisted on the proper use of gloves and sinks, those six accidental infections would not likely have occurred (Mermel et al. 1997).

Needle sticks have continued to be problematic in laboratories. Over the past decade, there have been multiple significant needle stick incidents. For instance, in 2004, a researcher at USAMRIID received a needle prick in biosafety level 4 containment while using a syringe on mice infected with a mouse-adapted variant of Ebola Zaire (Kortepeter et al. 2008). Also in 2004, a researcher at VECTOR in Russia died after pricking herself with a needle laden with the Ebola virus (Miller 2004). A researcher at the University of New Mexico was reportedly "jabbed with an anthrax-laden needle" in 2004 (The Sunshine Project 2007). A worker at the University of Chicago in 2005 reportedly "punctured his or her skin with an infected instrument bearing a BSL-3 select agent. It was likely a needle contaminated with either anthrax or plague" (The Sunshine Project 2007). A researcher at the Bernard Nocht Institute for Tropical Medicine in Hamburg, Germany, accidently pricked herself with a needle while working with Ebola virus in 2009 (The Canadian Press 2009).

Recent lapses in containment have also led to the introduction of disease into the community beyond the laboratory facility. In 2000, in Vladivostok, Russia, eight children ages 11–14 became ill after playing with discarded smallpox vaccine vials. The cause was most likely improper decontamination and disposal procedures by a nearby public health station (Byers 2009). Two laboratory workers in the National

Institute of Virology in Beijing contracted severe acute respiratory syndrome (SARS) between March and May 2004 after working with improperly inactivated virus. This led to seven additional people being infected in the community (US Centers for Disease Control and Prevention 2004). In August 2007, foot and mouth disease virus was released into the environment from a laboratory in the village of Pirbright in the United Kingdom, resulting in a significant local outbreak of the disease. The most likely cause of this release was determined to be improper liquid waste disposal as a result of the deteriorated condition of the site drainage system (Health and Safety Executive 2007).

There were also several notable security incidents at major bioscience facilities during this same time period. The US Federal Bureau of Investigations accused Bruce Ivins, a researcher at USAMRIID, of manipulating and distributing anthrax to cause harm. Specifically, the FBI claimed that, in 2001, Ivins mailed several letters that contained anthrax spores through the US Postal Service to various recipients across the United States, resulting in the deaths of 5 people and the sickening of 17 others (Federal Bureau of Investigation 2011). In 2004, Texas Tech University professor Thomas Butler was sentenced to two years in jail and multiple fines after he reported that 30 vials of plague bacteria were missing from his laboratory. After setting off a bioterrorism scare, he signed a statement that he accidentally destroyed the samples during the cleanup of a prior accident in his laboratory. However, it remains unclear what happened to those samples. Butler later recanted his signed statement and indicated it was possible that the samples had been destroyed, but he could not remember (Tanne 2003). During a general inspection of Texas A&M University in 2007, the CDC cited the university for failure to inform the CDC's Division of Select Agents and Toxins of a series of restricted aerosolization experiments with *Coxiella burnetii* on nine occasions from May 2003 to June 2005 (US Centers for Disease Control and Prevention, DSAT 2007). A former researcher at the National Microbiology Laboratory in Winnipeg, Canada, stole 22 vials of Ebola virus genetic material, which was discovered as he attempted to cross the US-Canada border in May 2009 (CBC News 2009). There is also a series of documented inventory discrepancies in a variety of US laboratories (US Centers for Disease Control and Prevention, DSAT 2007; Palk 2009; Sherman 2009; Margolin and Sherman 2005).

Recognizing the spate of safety and security incidents in bioscience facilities that had occurred in the late 1990s and the early 2000s, the editors of the fifth edition of the BMBL, which was published in 2007, emphasized the need for more comprehensive risk assessments in the field. It coached practitioners to assess the risk based on the agent hazards, to consider the hazards from the specific laboratory procedures, and then to "make a final determination of the appropriate biosafety level and select additional precautions based on the risk" (US Centers for Disease Control and Prevention, DSAT 2007). However, this guidance remained embedded in the historical paradigm of biosafety levels and agent risk groups.

A MODEL OF BIORISK MANAGEMENT BEGINS TO EMERGE

The following year a bipartisan US congressional commission released the *World at Risk* report, which among many recommendations called for bioscience laboratories that handle dangerous pathogens to implement a unified laboratory biorisk management framework to enhance their safety and security (Commission on the Prevention of WMD Proliferation and Terrorism 2008). Prior to the publication of this report, a group of experts from 24 different countries gathered to reconsider the traditional biosafety paradigm. In 2008, the European Committee for Standardization published a workshop agreement on laboratory biorisk management that reflected an emphasis on rigorous and experiment-specific and facility-specific risk assessment and mitigation, and management monitoring of performance with an emphasis on continual improvement. This document, known as CWA 15793, dismissed the conventional approach based on biosafety levels, received wide acclaim internationally, and was renewed in 2011 (European Committee for Standardization 2011). No other document of its kind exists in the international community.

Since its publication, many institutions around the world have initiated the process to implement CWA 15793 in order to better manage biorisks in their facilities. However, many organizations in North America, including some governmental agencies, remain skeptical of the value of biorisk management, and interpret it as simply an additional financial burden on the life sciences community (Steenhuysen and Begley 2014). At the time of this writing, the International Organization for Standardization (ISO) has initiated a new work item proposal on laboratory biorisk management. If approved by the member states, this project will pursue the development of an ISO international standard in biorisk management—the first of its kind in the biosciences.

Despite the publication of CWA 15793, biological scientists and facilities in the United States have been slow to embrace biorisk management—a culture of rigorously assessing risks, deciding how to mitigate those risks deemed to be unacceptable, and establishing mechanisms to constantly evaluate the effectiveness of the control measures. In 2013, the University of California–Los Angeles released initial results from its landmark international survey of laboratory safety. Almost half of the 2,400 scientists who responded had experienced injuries ranging from animal bites to chemical or biological inhalation. Thirty percent of respondents said they had witnessed at least one major injury that required attention from medical professionals. Perhaps most interesting was the discrepancy between US and UK scientists related to the use of risk assessments. In the United Kingdom, where the Health and Safety Executive mandates risk assessments, almost two-thirds of scientists said that they regularly execute risk assessments. In the United States, only one-quarter of scientists acknowledge that they conduct formal risk assessments; more than half of US scientists said they assessed risk only "informally" (Van Noorden 2013). Clearly, biorisk management is not yet embedded in the bioscience mindset of the United States.

Unfortunately, and perhaps not surprisingly, high-profile laboratory accidents remain somewhat commonplace—even in some of the world's most sophisticated bioscience facilities. In 2012, the CDC reported that there were 727 incidents of theft, loss, or release of select agents in the United States between 2004 and 2010, resulting in 11 laboratory-acquired infections (Henkel et al. 2012). In early 2014, a laboratory at the US Centers for Disease Control and Prevention in Atlanta, Georgia accidentally and unknowingly cross-contaminated low-pathogenic influenza samples with the highly pathogenic H5N1 influenza, and transferred the material to laboratories that were not approved to work with H5N1. The CDC did not learn of the mistake until informed by one of the facilities that had received the sample. Later in the same year, as many as 84 workers were unknowingly and unintentionally exposed to live strains of the *Bacillus anthracis* at the CDC in Atlanta (Russ and Steenhuysen 2014). Scientists in a BSL3 laboratory failed to inactivate the bacteria, and neglected to validate the inactivation, before transferring the material to three BSL2 laboratories.

NEW FOCUS ON RISKS OF BIOSCIENCE RESEARCH

This less-than-stellar safety and security record has made the general public, and even some in the scientific community, question the fundamental rationale for work with dangerous biological agents. In 2003, Boston University Medical Center won a grant from the National Institutes of Health to build one of two national biocontainment laboratories—the National Emerging Infectious Disease Laboratory—as part of the new US biodefense research strategy. Yet in 2014, because of the protracted public opposition, research had not begun in the biosafety level 4 suites, and the Boston mayor sought an ordinance "to ban level 4 research as proper precaution in light of the possibility that safeguards might fail" (Boston Globe Editorial 2014). In 2011, researchers at the Erasmus Medical Center in Rotterdam, the Netherlands, led by Ron Fouchier, and the University of Wisconsin–Madison, led by Yoshihiro Kawaoka, in separate studies artificially engineered the H5N1 avian flu virus to transmit easily from one ferret to another (Herfst et al. 2012; Imai et al. 2012). Since this research used an established animal model for human flu, it effectively created a potentially pandemic strain of influenza that does not currently exist in nature. In 2014, the Wisconsin team combined the genes from several avian flu viruses to construct a new organism similar to the 1918 Spanish flu virus that also spread efficiently in ferrets (Watanabe et al. 2014). Many scientists have argued that the risk of accidental release, accidental infection, or intentional misuse of this so-called gain of function research does not outweigh the scientific benefits of increased knowledge about how avian influenza might naturally and genetically evolve into something particularly more dangerous. Based on historical laboratory-associated infections in BSL3 facilities, Marc Lipsitch and Alison Galvani concluded that over a 10-year period there would be a 20% risk of at least one laboratory-acquired infection of a novel pandemic flu strain, which could initiate an extensive spread of the disease (Lipsitch and Galvani 2014). Although there are fierce arguments on both sides regarding the legitimacy of gain of function research, the fundamental concern about this issue hinges on laboratory biosafety and biosecurity.

The bioscience community has been acutely aware of the risks associated with biological research, specifically that it could be misused for malevolent purposes, or that it could result in the creation of novel pathogens with unique properties—perhaps even an entirely new class of threat agents. In 2004, the National Academy of Sciences published the so-called Fink Report, which defined seven categories of experiments of concern, and developed a series of recommendations to prevent the misuse of biology without preventing the conduct of legitimate research (National Research Council 2004). Unfortunately, the Fink Report did not identify the need

to improve the practice of laboratory biosafety and biosecurity, other than to rec-
ommend "that the federal government rely on implementation of current legislation
and regulation, with periodic review by the National Science Advisory Board on
Biosecurity (NSABB), to provide protection of biological materials and supervision of personnel working with these materials" (National Research Council 2004).
However, the NSABB has conducted no substantive review of laboratory biosafety
and biosecurity since it was established in 2005.

The rapid advance of synthetic biology further highlights the critical need to
reevaluate the current biosafety and biosecurity system based on agent risk groups,
biosafety levels, and security regulations. Leveraging significant, recent improvements in the ability to synthesize and sequence DNA, synthetic biology pursues the
creation of technologies for designing and building biological organisms—and often
these organisms are completely novel or have unique characteristics. These new biological agents obviously do not appear on existing agent risk groups or select agent
lists, or fall neatly into traditional biosafety levels. Arguably and understandably,
the field of bioethics has become much more robust with the emergence of synthetic
biology: traditional methods of biosafety and biosecurity seem largely irrelevant, so
the public must increasingly rely on the ethical behavior of the scientists involved
to ensure safety and security. Notably, the synthetic biology field has recently begun
recognizing the need to develop more robust approaches to risk and risk assessment (Pauwels et al. 2013). A study by the Netherlands Commission on Genetic
Modification (COGEM) in 2013 concluded that current risk assessment approaches
may be sufficient for synthetic biology today, but will not be sufficient to address
risks when there is no known reference organism or introduced characteristics are
unpredictable (Commission on Genetic Modification 2013). In 2010, a National
Academy of Sciences report concluded that a "sequence-based prediction system for
oversight of Select Agents is not possible now and will not be possible in the usefully
near future" (National Research Council 2010).

Clearly, the rapid technological advances of the biosciences compel the community to reconsider the traditional methods of ensuring safety and security. The
power of biotechnology to counter the threat that emerging infectious diseases pose
to public and economic health, and global accessibility to this technology, has also
led to a rapid expansion of sophisticated laboratories around the world (Fonkwo
2008). The risk of a catastrophic biosafety or biosecurity incident seems to increase
on an almost daily basis, especially if the traditional, rule-based biosafety paradigm
remains unchanged. Today, the bioscience community needs to develop and adopt a
new, performance-based method to manage the risks of the biosciences—before it
is responsible for a major catastrophe. This book argues that biorisk management is
the solution.

LEARNING LESSONS FROM OTHER INDUSTRIES

More than half a century ago, in 1961, G.B. Phillips noted that "in a broad sense,
attitudes and activities which create conditions favorable for occupational infections
are similar to those that lead to the occurrence of industrial type accidents" (Phillips

1961). We also believe that much can be learned from understanding how different industries approach safety and security. In particular, those industries that have experienced major accidents, involving large-scale loss of life, have been compelled to reassess their safety programs, and have almost universally recognized deficiencies in risk management as a principal cause in those accidents. As a result, those industries have embraced a performance-based, holistic, risk management paradigm. Generic, rule-based, administrative systems have been abandoned. The bioscience community should not wait for a calamity to occur in its field before learning these fundamental lessons. The following section describes a number of catastrophic accidents that have compelled widely different industries—outside of the life sciences community—to embrace broad-ranging risk management systems. It also demonstrates that industries that have implemented substantive risk management policies, standards, and expectations have dramatically decreased both the number and severity of accidents.

UNION CARBIDE, BHOPAL, 1984

On December 3, 1984, a large toxic vapor cloud containing 40 tons of methyl isocyanate (MIC) gas leaked from a Union Carbide pesticide plant in Bhopal, India. Prevailing winds carried the chemical cloud over the city of Bhopal and exposed well over 500,000 people. The numbers of reported fatalities and injuries vary widely, but recent studies indicate that the MIC contamination of Bhopal killed over 10,000 people, caused as many as 20,000 premature deaths, and injured or disabled another 50,000 (Broughton 2005). This accident is still considered the worst chemical plant disaster in industrial history.

The leak was caused by water being misdirected to one of the two large MIC storage tanks during a routine pipework cleaning activity. MIC mixed with water causes an exothermic reaction, and resulted in a large volume of a toxic mixture forming and eventually being released into the atmosphere through the plant's flare tower. There were a number of safeguards at the plant designed to prevent this sort of release, but almost all of these preventative measures were inactive or not monitored (American University 2014; Manaan 2005).

The Bhopal disaster was a classic example of egregious mismanagement at every level of the organization, from executive management to the individual tank operators. For example, the operators who saw the increased pressure on the MIC tank pressure gauges thought the poorly maintained gauges were giving false readings. Although the flare tower was designed to flare off any vented vapors from the plant, it did not function properly. At the time of the accident, Union Carbide could not provide specific details of the effects of MIC and the immediate treatment required for those exposed. The company had no emergency response plan in place. The plant management team, all of whom were Indian nationals, was sentenced to two years in jail, and the CEO of UC Corporation fled the country after being released on bail (Karasek 2014).

The most notable positive legacy from Bhopal was the widespread adoption of process safety as a professional engineering discipline, and as a requirement throughout the chemical industry worldwide. In 1988, the American Chemistry Council started to implement a program originally developed in Canada called Responsible Care to reduce the potential chemical risks to workers and the environment. Three years later the International Council of Chemical Trade Associations (ICCTA) was founded, with one of its principal objectives being the promotion of Responsible Care practices worldwide. The Responsible Care program now extends to 47 countries, corresponding to 85% of the world's chemical producers. In 1990, with the passage of the Clean Air Act Amendments, the US Occupational Safety and Health Administration (OSHA) was directed to create and enforce the "Process Safety Management of Highly Hazardous Chemicals" (PSM). This PSM standard is a comprehensive program that integrates technologies, procedures, and management practices (Manaan et al. 2005).

CHERNOBYL NUCLEAR POWER PLANT, 1986

On April 26, 1986, a sudden power output surge occurred during an unauthorized systems test at the Chernobyl Nuclear Power Plant near the town of Pripyat, Ukraine (then part of the Soviet Union). During an attempted emergency shutdown, the reactor vessel ruptured, a radioactive fire began, and a plume of radioactive fallout was sent into the atmosphere. The plume of primarily iodine and cesium radionuclides drifted over large parts of the western Soviet Union and much of Europe. Approximately 150,000 square kilometers were contaminated, affecting more than 5 million people in Belarus, Russia, and Ukraine. It is estimated that over 330,000 people were evacuated and resettled from the most severely contaminated areas. About 240,000 workers took part in the recovery, operating within the 30-kilometer zone surrounding the reactor that is now uninhabited. More than 30 reactor staff and emergency workers immediately died from the radiation exposure, and as many as 4,000 people ultimately died from the accident (International Atomic Energy Agency 2006).

The Chernobyl accident was the result of a flawed reactor design that was operated with inadequately trained personnel and the absence of any safety culture. The design of the plant placed a heavy dependence on adherence to administrative controls and procedures for safe operation. However, there was very little training of the plant operators on what these administrative and procedural controls were, and the safety implications for failing to implement these controls (International Nuclear Safety Advisory Group 1992; Engineering Failures 2009).

Following the Chernobyl disaster, significant safety design modifications were made to all nuclear reactors similar in type to those operating in Chernobyl. Automatic shutdown mechanisms now operated faster, and other safety mechanisms were improved. Automated inspection equipment was also installed. In addition, safety training for reactor workers dramatically increased, as well as a heightened focus on operational and management systems and regulatory oversight. The US nuclear industry instigated a self-policing agency called the Institute of Nuclear Power Operations with a mission of promoting the "highest level of safety and reliability" (Karasek 2014). In addition, the US Nuclear Regulatory Commission began evaluating the role of risk assessment, management systems, and performance-based indicators and regulatory frameworks (Walker and Wellock 2010).

PIPER ALPHA OIL PLATFORM, 1988

On July 6, 1988, a gas explosion occurred on the Piper Alpha Oil Production Platform, operated by the Occidental Group and located 120 miles northeast of Aberdeen, Scotland, in the North Sea. In 22 minutes, the subsequent fire killed 167 of the 228 people on board the oil platform. The Piper Alpha extracted oil from beneath the ocean floor, and processed natural gas. It served as a hub, connecting the gas lines of two other Piper field platforms. At the time of the accident, Piper Alpha exported just under 120,000 barrels of oil and approximately 33 million standard cubic feet of gas per day (National Aeronautics and Space Administration 2013).

On the day of the accident, a worker removed the pressure safety valve on the Piper Alpha's Pump A during routine maintenance and replaced it with a round metal plate—called a blind flange. Since he was not able to finish the maintenance, he completed a form stating that Pump A was not ready for operation and should not be activated. During a subsequent shift later that night, Pump B failed, halting all offshore production unless Pump A could be restarted. Unable to find the mainte- nance documentation, and believing that Pump A was safe to use, workers activated Pump A, which caused a high-pressure gas leak and explosion. Gas production from the two other Piper field platforms was not shut down, providing continuous fuel to a massive fire on Piper Alpha (Karasek 2014).

The escape of gas from Pump A sparked the initial explosion, but the fateful deci- sion to activate Pump A was caused by the absence of a reliable process to ensure that workers documented and communicated from shift to shift the maintenance sta- tus and operability of technical systems. Moreover, the absence of blast walls on the Piper Alpha platform and the failure to shut down oil and gas supplies from the two other Piper field platforms prevented any containment of the initial fire and fueled a much larger conflagration (Scott 2011).

The official investigation of the disaster, known as the Lord Cullen Report (Cullen 1990), resulted in 106 recommendations for changes to North Sea safety procedures—all of which were accepted by the industry, including the formation of a new UK government organization, the Health and Safety Executive (HSE). In a speech commemorating the 25th anniversary of the tragedy, Lord Cullen said, "I dis- covered it was not just a matter of technical or human failure. As is often the case, such failures are indicators of underlying weaknesses in management of safety" (Oil and Gas Industry Association 2013). In addition to many technical and infrastructure failures, training, monitoring, and auditing had been poor, the lessons from a previ- ous relevant accident had not been learned, and evacuation procedures had not been practiced. Perhaps most importantly, Occidental had not conducted a risk assessment of the major hazards, and determined how to control them. According to Cullen, "The quality of management safety is fundamental and that depends critically on effective safe leadership at all levels and the commitment of the whole workforce to give priority to safety" (Harris 2013).

TEXAS CITY REFINERY, 2005

On March 23, 2005, an explosion and fire occurred at BP Products North America, owned and operated by Texas City Refinery, killing 15 and injuring 180 people, and resulting in financial losses exceeding $1.5 billion. The Texas City Refinery was the third largest oil refinery in the United States, with an input capacity of 437,000 barrels per day as of January 2000 (Wikipedia 2014).

The incident occurred after a release of a flammable liquid geyser from a blowdown stack that was not equipped with a flare, leading to an explosion and fire. According to BP's own accident investigation report, the cause of the accident was "heavier-than-air hydrocarbon vapors combusting after coming into contact with an ignition source, probably a running vehicle engine" (British Petroleum 2005).

According to the US Chemical Safety and Hazard Investigative Board, the disaster was caused by both technical mistakes and failures as well as "organizational and safety deficiencies at all levels of the BP Corporation." In particular, the BP Board of Directors did not provide effective oversight of the company's safety culture and major accident prevention programs; the plant management mistakenly believed that a low personal injury rate was an accurate indicator of process safety performance; the mechanical integrity program resulted in a "run to failure" of process equipment; the safety policy and procedural requirements were operated with a "check the box" mentality; the plant lacked a safety reporting and learning culture; and safety campaigns and goals focused on improving personal safety metrics rather than process safety and management safety systems (US Chemical Safety and Hazard Investigation Board 2007).

BP subsequently commissioned an independent report to evaluate BP's corporate safety management systems, safety culture, and oversight of the North American refineries. Known as the Baker Panel Report, it found that an "apparent complacency toward serious process safety risk existed at each refinery," and BP's corporate safety management system "does not ensure adequate identification and rigorous analysis of risks" and "does not effectively measure and monitor process safety performance." The first two recommendations in the report included providing "process safety leadership" at the highest levels of BP executive management, and establishing and implementing an "integrated and comprehensive process safety management system" (Baker et al. 2007).

FUKUSHIMA NUCLEAR POWER PLANT, 2011

On March 11, 2011, a magnitude 9.0 earthquake struck the eastern coast of Japan, triggering a tsunami that flooded 560 square kilometers along the coast under several meters of water. The earthquake caused the failure of the safety systems at the Fukushima Daiichi Nuclear Power Plant, owned and operated by the Tokyo Electric Power Company (TEPCO). The resulting release of radioactive material was eventually declared a Level 7 "severe accident" by the International Nuclear Event Scale—one of only two Level 7 incidents in history. Experts estimate that 900,000 terabecquerels of radioactive material were released during the disaster (Reuters 2012). One hundred sixty thousand people were evacuated from the area around Fukushima; 600 died during the evacuation (McGreal 2012). The area around the plant remains severely radioactive, and will be uninhabitable for decades. Estimates of the total economic loss as a result of the Fukushima disaster range from US$250 billion to $500 billion (Starr 2012). According to a recent study by Stanford University scientists, the fallout from Fukushima will most likely cause 180 additional cancers and 130 deaths, most of them in Japan (Ten Hoeve and Jacobson 2012).

The report of the Japanese government's independent investigative commission cataloged "a multitude or errors and willful negligence that left the Fukushima plant unprepared," and concluded that Fukushima "was a profoundly manmade disaster—that could and should have been foreseen and prevented. And its effects could have been mitigated by a more effective human response." The commission did not criticize the competency of any specific individuals, but argued that the fundamental causes were "the organizational and regulatory systems that supported faulty rationales for decisions and actions." In particular, the report accused the government, the regulators, and TEPCO of colluding with one another "based on the organization's self interest, and not in the interest of public safety." The independence of the regulators "from the political arena, the ministries promoting nuclear energy, and the operators was a mockery." Criticizing the insularity of Japanese culture, the report accused the regulators of "a negative attitude toward the importation of new advances in knowledge and technology from overseas" (National Diet of Japan 2012).

TEPCO, the plant's operator, received the harshest criticism: "TEPCO did not fulfill its responsibilities as a private corporation.... The risk management practices of TEPCO illustrate this." Although TEPCO was aware of the risk that a tsunami could result in a total outage of electricity at the plant, and that the loss of seawater pumps could damage the reactor cores, TEPCO chose not to implement any measures to reduce or eliminate these risks. TEPCO was only concerned with the "risk to their own operations." In addition to accepting these catastrophic risks, TEPCO had no response measures in place for any severe accident: "no manual or training regimens." Instead, TEPCO subordinated itself to the Prime Minister's Office during the emergency response, avoided transparency, and was reluctant to take responsibility. Among its recommendations, the report asserted that "TEPCO must undergo dramatic corporate reform, including governance and risk management and information disclosure—with safety as the sole priority" (National Diet of Japan 2012).

Y-12 NUCLEAR FACILITY, 2012

On July 28, 2012, three protestors broke into the Y-12 nuclear weapons production facility in Oak Ridge, Tennessee. After using simple tools to cut through four alarmed fences, the protestors defaced the outside of the Highly Enriched Uranium Materials Facility (HEUMF) without being interrupted by the security measures in place. Eventually, a protective force officer was dispatched to assess the alarms, but he did not notice the protestors until they approached his vehicle and "surrendered." Fortunately, these protestors were not terrorists who intended to commit an act of nuclear sabotage or steal highly enriched uranium for use in a nuclear weapon.

An investigation of the incident by the US Department of Energy's Inspector General (DOE IG) revealed "multiple system failures on several levels. For example, we identified troubling displays of ineptitude in responding to alarms, failures to maintain critical security equipment, over reliance on compensatory measures, misunderstanding of security protocols, poor communications, and weaknesses in contract and resource management" (US Department of Energy 2012). The site had a new intrusion detection system that had an unusually high rate of false alarms. The guards would normally respond to an alarm by using a camera to verify if there was an intruder. But because the cameras had been broken for months, without a clear plan for their repair, guards were sent to check on the alarms. However, the guards had grown complacent and weary of investigating false alarms. Poor communications led officers to incorrectly assume the protestors hammering on the wall of the HEUMF were plant maintenance workers. The officer who eventually responded to the situation did not immediately secure the area or detain the protestors. The DOE IG investigation suggests that the facility had not adequately considered the risk of unarmed protestors assaulting the site, and was not prepared to respond to such an event. According to the report, "the actions of these officers were inconsistent with the gravity of the situation and the existing protocols" (US Department of Energy 2012).

Nuclear security in the United States is heavily regulated and highly scrutinized. Despite all the rules and regulations, and the resources dedicated to the protection of nuclear materials, technologies, and expertise, there was a fundamental breakdown in the management and oversight of the security system—the absence of a robust security risk management culture. A Harvard University report on nuclear security noted that the primary lesson to be learned from the Y-12 incident is: "People and organizations matter—a poor security culture can severely undermine security even at facilities with modern security equipment, extensive security spending, stringent security rules, and regular security testing" (Bunn et al. 2014). Responding to the Inspector General's report, Thomas D'Agostino, administrator of the National Nuclear Security Administration, called the incident a "wake-up call for our entire complex," and pledged to make a series of "structural and cultural changes" to improve existing nuclear security risk management systems (US Department of Energy 2012).

PHARMACEUTICAL INDUSTRY

A series of crises in the pharmaceutical industry in the 20th century resulted in the creation of a substantive regulatory environment for the production, testing, and sale of pharmaceutical products internationally. In June 1937, the SE Massengill Company distributed 633 shipments of the elixir sulfanilamide in a liquid form throughout the United States. Used in powder and tablet form, sulfanilamide had been prescribed to treat streptococcal infections. To make the liquid form, the company dissolved the sulfanilamide in diethylene glycol—a deadly poison. At the time, the food and drug laws did not require pharmacological studies on new drugs. In September 1937, liquid sulfanilamide caused the deaths of 107 people across 15 states. Following this incident, laws were immediately enacted that required manufacturers to show that a drug was safe before it could be marketed (Ballentine 1981; Meadows 2006a).

In the late 1950s, thalidomide was marketed in Europe, Australia, and Japan as a sedative, and was used in the treatment of nausea in pregnant women. Within a few years of the widespread use of thalidomide, approximately 10,000 children were born with phocomelia (Lenz 1988). In addition to limb reduction anomalies, other effects later attributed to thalidomide included congenital heart disease, malformations of the inner and outer ear, and ocular abnormalities (Miller and Strömland 1999). By 1961, most countries had banned the sale and use of thalidomide, and laws were soon passed that required drug firms to prove both safety and effectiveness for any product's intended use (Braithwaite and Drahos 2000).

In 1982, for reasons not known, a malevolent person or group replaced Tylenol Extra-Strength capsules with cyanide-laced capsules, resealed the packages, and placed them on the shelves of several pharmacies and food stores in the Chicago area. The poison capsules were purchased, and seven unsuspecting people died a horrible death. Soon after, laws were passed that made it a crime to tamper with packaged consumer products (ten Berge 1990; Meadows 2006a).

The World Health Organization introduced the first good manufacturing practices (GMP) guidelines for manufacturing, processing, packing, or holding finished pharmaceuticals in the 1960s, and many countries then developed their own GMP guidelines based on the WHO guidelines. In 1975, the World Health Organization began internationalizing pharmaceutical regulation by requiring countries that export pharmaceuticals to certify those companies as manufacturers of drugs that are authorized for the domestic market and with production facilities that are regularly checked for GMP compliance (World Health Organization 2003; Brhlikova et al. 2007).

The introduction of GMP rules served as a great leap forward for the pharmaceutical industry with regard to product quality and safety. However, the GMP process itself remained a reactive regulatory process. Revised GMP guidelines were frequently released in the 1970s and 1980s to account for every new issue or perceived problem in the system. Product labeling and cleaning validation are examples of this reactive regulatory approach.

One of the drawbacks to the proliferation of GMP rules and regulations was the divergence in technical requirements from country to country at a time when the pharmaceutical industry was becoming increasingly international. In 1990, the International Conference on Harmonisation of Technical Requirements for Registration of Pharmaceuticals for Human Use (ICH) was formed to help lower the costs of healthcare and research and development, and minimize the delay in making new and efficacious treatments available to the public (ICH 2014).

National legislation in the early 1990s also reflected the desire to allow patients faster access to promising therapies, to create a more efficient regulatory process for industry to market new drugs more quickly, and to increase the revenues and profits of the industry. By the late 1990s, concern about the increased pace of drug approvals had unintentionally led to a neglect of safety considerations. In fact, there was a spate of drug withdrawals in the late 1990s that contributed to a public perception that the drug safety system was in crisis (Friedman et al. 1999; Institute of Medicine 2006).

GMP rules and regulations, which had primarily focused on technical and physical control measures, had not solved all the drug safety challenges. Adopting lessons from other industries, the pharmaceutical industry recognized that it needed to adopt a risk-based approach to drug safety. Rather than focusing on safety and efficacy only at the initial drug approval stage, the industry needed to sustain active reassessments of risks as drugs entered and were used by a growing number and diversity of patients. By 2005, the ICH published ICH Q9, "Quality Risk Management," which, according to the Parenteral Drug Association, "is a systematic process for the assessment, control, communication, and review of risks to quality of the drug product across the product lifecycle" (O'Mahony 2011). The implementation of quality risk management has been frequently cited as responsible for a significant reduction in the number of drug withdrawals and an end to the drug safety crisis at the turn of the century.

FOOD INDUSTRY

The food production sector has also experienced many serious incidents that have compelled that industry to implement modern risk management systems and global standards. Two catastrophes, in particular, shook the food production industry to its core.

In 1986, cattle in the United Kingdom began to suffer from a condition that was described as mad cow disease, and by 1987 the British Ministry of Agriculture acknowledged that the country was besieged by an outbreak of bovine spongiform encephalopathy (BSE). It was estimated that over 460,000 cattle infected with BSE entered the human food chain in the 1980s as a result of cattle being fed the remains of other cattle in the form of meat and bone meal. In 1989, the United States prohibited the import of live cattle, sheep, bison, and goats from countries where BSE was known to exist. Finally, in March 1996, the UK government acknowledged that BSE could be transmitted to humans and cause a variant form of Creutzfeldt–Jakob disease (vCJD). Immediately, the British destroyed 4.5 million cattle, and the European Union imposed a ban on the import of British beef that lasted more than 10 years. Nevertheless, over 200 humans have died from vCJD since the outbreak, and evidence indicates that they had consumed tainted beef (Valleron et al. 2001; University of Edinburgh 2009; Center for Food Safety 2014).

In 1993, the Jack in the Box restaurants in the United States began a special promotion of the Monster Burger, selling the sandwich at a deep discount. The advertising campaign was so successful that the demand for that particular burger overwhelmed the restaurants. Ultimately, 73 different Jack in the Box restaurants in the western United States sold undercooked hamburgers that were infected with what was then a little known bacterium called *E. coli* O157:H7, and more than 730 people became ill. Four children died and 178 other victims suffered permanent injury, including kidney and brain damage (US Centers for Disease Control and Prevention 1993; Benedict 2011; Manning 2010). Although there had previously been 22 documented outbreaks of *E. coli* O157:H7 in the United States, resulting in 35 deaths, the Jack in the Box outbreak sparked such wide-scale and dramatic media coverage that Senator Richard Durbin described it in 2006 as a "pivotal moment in the history of the beef industry" (US Senate 2006).

These two major events, along with many other food contamination incidents around this time, compelled the food industry to reexamine its own processes and standards. *E. coli* became a reportable disease for all state departments of health, and the US Department of Agriculture reclassified *E. coli* O157:H7 as an adulterant in ground beef, and began regularly testing for it. Inspired by actions taken by Jack in the Box immediately following the outbreak, the US Food and Drug Administration established the Hazard Analysis and Critical Control Points (HACCP) program. The first HACCP regulations went into effect in 1997.* HACCP is a science-based initiative that involves identifying potential hazards and risks, monitoring targeted critical control points, and recording and reviewing results. It is a proactive management system designed to prevent hazards rather than a reactive process of responding to contaminants (Meadows 2006b).

Shortly thereafter, the Consumer Goods Forum, the only independent global network for consumer goods retailers and manufacturers worldwide, launched the Global Food Safety Initiative (GFSI) to spearhead continuous improvement of food safety management systems around the world. GFSI defines food safety requirements along the entire food supply chain to cover issues such as feed, distribution, and packaging, and helps set international standards and expectations.† Today, there is a series of international standards for food safety management, such as the ISO 22000 family of standards, which help organizations identify and control food safety hazards, and implement comprehensive food safety management systems (International Organization for Standardization 2005).

At least partly as a result of these industry and regulatory efforts to establish a risk management culture and system, the occurrence of foodborne pathogens causing illnesses in the United States dropped significantly between 1999 and 2010. According to a 2010 report by the Centers for Disease Control and Prevention and data from the CDC's Foodborne Disease Active Surveillance Network (FoodNet), rates of infection from food contaminated with one of the five primary bacterial pathogens declined at least 25% over the previous decade (Osterholm 2011; Scallan et al. 2011).

AIRLINE INDUSTRY

Today, aviation is recognized around the world as one of the safest means of transport. According to the International Civil Aviation Organization (ICAO), the safety performance of the aviation industry has improved by a factor of more than 130 times over the course of the past 60 years (Graham 2010). And the most dramatic improvement has come in the past decade or so. The year 2012 was the safest since 1945, with only 475 fatalities worldwide—fewer than half the 1,147 deaths in 2000 (Mouawad and Drew 2013).

Clearly, the airline industry has recognized that its success must be built on a foundation of constantly improving safety. The outstanding track record for aviation

---

* In 1995, the FDA established the Seafood HACCP, which went into effect in 1997. The FDA developed a HACCP program for meat and poultry processing plants in 1998. The FED established the Juice HACCP regulations in 2001, and they went into effect in 2003 (US Food and Drug Administration 2014).

† The Global Food Safety Initiative was launched in 2000 (Global Food Safety Initiative 2014).

safety is a result of an industry that remains on the cutting edge of safety management issues. In particular, the aviation industry was among the first major industries to define organizational safety, and to embrace risk management systems as an essential component to reduce the frequency and impact of aviation accidents.

The International Civil Aviation Organization first published its groundbreaking *Safety Management Manual* (SMM) in 2003 after years of studying aviation accidents and aviation safety. ICAO published the third edition of the SMM in 2013. The SMM describes three historical eras of progress in aviation safety (International Civil Aviation Organization 2013).

Beginning in the early 1900s until the late 1960s, when aviation emerged as a form of mass transportation, aviation accidents and disasters were related to technical factors and technological failures. As a result, the focus of safety efforts was on post-accident investigations and technological improvements, and eventually regulatory compliance and oversight that also focused on technical components of aviation safety. Major technological advances and improved safety regulations contributed to a significant reduction of the frequency of aviation accidents during this period.

From the early 1970s until the mid-1990s, aviation safety experts recognized that accidents occurred not simply because of poor technology or lack of regulatory oversight, but failures in human performance. The application of human factors science focused on the role of the individual and his or her understanding and application of the technical safety components. Specifically, aviation safety emphasized how to integrate human operators and their behaviors most safely and efficiently with the technology.

Beginning in the mid-1990s, aviation safety experts began to view the issue from a systemic perspective, acknowledging that organizational factors contributed as much or more than human and technical factors to aviation safety. The notion of the organizational accident was introduced, and the influence of organizational culture and policies on the effectiveness of risk controls was recognized. Traditional collection and analysis of accident data was supplemented by a proactive approach that assessed and monitored safety risks, and identified emerging safety issues. Safety management systems were developed specifically to mitigate the risks of organizational accidents by reducing their likelihood and severity.

In 2006, ICAO began mandating that each of its member states implement a safety program that established safety targets and tracked performance in reaching those objectives. At the same time, ICAO required airlines, airports, air traffic operators, and aircraft maintenance organizations to create safety management systems to identify and assess hazards and risks on a continuing basis and to apply risk mitigating measures before accidents occurred (International Civil Aviation Organization 2007). These risk management systems principles were quickly embraced by international nongovernmental aviation safety groups, and reflected in new governmental regulatory initiatives.*

---

* The Commercial Aviation Safety Team is the leading international aviation safety alliance (Commercial Aviation Safety Team 2014). The US Federal Aviation Administration introduced a safety management system regulation for commercial airlines in 2010 (US Federal Aviation Administration 2010).

BIORISK MANAGEMENT

Clearly, risk management systems are common across many industries, especially those industries in which accidents can have significant consequences. The root cause of the high-consequence accidents described in the previous section extended well beyond technical or human failure, and almost always reflected underlying failures in the *management* of safety or security. High-consequence industries that have implemented proactive risk management systems have achieved a significant reduction in the number and severity of accidents.

Despite the progress that many other industries have made to implement formal risk management systems, there is no universally accepted international system for biorisk management in the life sciences. This book aims to define that new paradigm.

Biorisk management aims to transform the traditional laboratory biosafety and biosecurity field, which is based on predetermined biosafety levels and prescriptive, checklist biosecurity regulations, and almost exclusively on technical and technology solutions and operator performance and behavior. With this traditional approach, a bias has persisted that facilities with more resources, or from countries or regions that are more developed, are by definition safer and more secure than facilities with fewer resources. Moreover, in the traditional model, safety and security almost always exist in two wholly separate silos in life science institutions, and risks to institutions are not evaluated and managed at the institutional level. As a result, these independent safety and security procedures are executed as an administrative function, and all responsibility is delegated down to the so-called biosafety officer, who is generally viewed (and compensated) as a monitor or regulator who often must operate in opposition to the scientific mission and the scientific staff.

By contrast, a biorisk management system emphasizes a depth of roles and responsibilities for everyone in the institution, and ensures that the highest levels of management have ultimate responsibility for the system. A risk management system places a priority on intellectually sound, evidence-based decision making. Risk assessments are substantive exercises that evaluate all of a facility's risks, and are based on the unique operations of the facility, not on generic agent risk statements or agent risk groups. Mitigation measures are implemented according to management's risk-based decisions, not based on a predetermined description of a biosafety level. Not only are risks evaluated proactively, but the performance of a biorisk management system is constantly assessed to help anticipate what could go wrong and how. The biorisk management system concept is explicitly scalable, applying to a research laboratory as well as a production facility, a hospital, or a field investigation. Moreover, because biorisk management is explicitly oriented around performance, an institution in a low-resource environment can implement biorisk management as effectively as an institution in the developed world.

Although new paradigms are never embraced easily or quickly, and every industry habitually adheres to conventional procedures and prior practices, we fervently believe that the rationale and concept of biorisk management is gaining strong momentum internationally. Clearly, the future for biorisk management is bright. We hope that this book can further define and contribute to the development of biorisk management.

REFERENCES

American University. 2014. Bhopal Disaster. TED Case Studies, Case 233. http://www1.american.edu/ted/bhopal.htm (accessed August 13, 2014).

Baker III, James A., F.L. Bowman, G. Erwin, S. Gorton, D. Hendershot, N. Leveson, S. Priest, I. Rosenthal, P.V. Tebo, D.A. Wegmann, and L.D. Wilson. 2007. The Report of the BP U.S. Refineries Independent Safety Review Panel. January. http://www.bp.com/liveassets/bp_internet/globalbp/globalbp_uk_english/SP/STAGING/local_assets/assets/pdfs/Baker_panel_report.pdf.

Ballentine, Carol. 1981. Taste of Raspberries, Taste of Death: The 1937 Elixir Sulfanilamide Incident. *FDA Consumer Magazine.* June. http://www.fda.gov/aboutfda/whatwedo/history/productregulation/sulfanilamidedisaster/default.htm.

Barbeito, Manuel S., and R.H. Kruse. 2014. A History of the American Biological Safety Association. http://www.absa.org/abohist1.html (accessed April 11, 2014).

Benedict, Jeff. 2011. *Poisoned: The True Story of the Deadly E. coli Outbreak That Changed the Way Americans Eat.* Buena Vista, VA: Inspire Books.

Boston Globe Editorial. 2014. 'Level 4' Disease Research Can Be Safe, Belongs in America's Medical Capital. *Boston Globe.* April 13. http://www.bostonglobe.com/opinion/editorials/2014/04/13/level-disease-research-can-safe-belongs-america-medical-capital/sShOiraz03EUmSRX9JCueO/story.html.

Braithwaite, John, and P. Drahos. 2000. *Global Business Regulation.* Cambridge, UK: Cambridge University Press.

Brhlikova, Petra, I. Harper, and A. Pollock. 2007. Good Manufacturing Practice in the Pharmaceutical Industry: Working Paper 3, Workshop on Tracing Pharmaceuticals in South Asia. University of Edinburgh. July 2–3. https://www.csas.ed.ac.uk/__data/assets/pdf_file/0011/38828/GMPinPharmaIndustry.pdf.

British Petroleum. 2005. Fatal Accident Investigation Report: Isomerization Unit Explosion, Texas City, Texas. December 9. http://www.bp.com/liveassets/bp_internet/us/bp_us_english/STAGING/local_assets/downloads/t/final_report.pdf.

Broughton, Edward. 2005. The Bhopal Disaster and Its Aftermath: A Review. *Environmental Health: A Global Access Science Source*, 4: 6.

Bunn, Matthew, M.B. Malin, N. Roth, and W.H. Tobey. 2014. Advancing Nuclear Security: Evaluating Progress and Setting New Goals. Harvard University, Kennedy School, Belfer Center for Science and International Affairs. March. http://belfercenter.ksg.harvard.edu/files/advancingnuclearsecurity.pdf.

Byers, K.B. 2009. Biosafety Tips. *Applied Biosafety*, 14(2): 99–102.

The Canadian Press. 2009. Canadian-Made Ebola Vaccine Used after German Lab Accident. http://www.cbc.ca/news/technology/canadian-made-ebola-vaccine-used-after-german-lab-accident-1.827949 (posted March 20, 2009).

CBC News. 2009. Winnipeg Researcher Charged with Smuggling Ebola Material into U.S. May 13. http://www.cbc.ca/news/canada/winnipeg-researcher-charged-with-smuggling-ebola-material-into-u-s-1.774725.

Center for Food Safety. 2014. Timeline of Mad Cow Disease Outbreaks. http://www.centerforfoodsafety.org/issues/1040/mad-cow-disease/timeline-mad-cow-disease-outbreaks (accessed April 2014).

Center for Global Development. 2014. Case 1: Eradicating Smallpox. http://www.cgdev.org/page/case-1-eradicating-smallpox (accessed August 12, 2014).

College of Physicians of Philadelphia. 2014. The History of Vaccines. http://www.historyofvaccines.org/content/timelines/smallpox (accessed August 12, 2014).

Commercial Aviation Safety Team. 2014. http://www.cast-safety.org (accessed April 2014).

Commission on Genetic Modification. 2013. Synthetic Biology—Update 2013: Anticipating Developments in Synthetic Biology. CGM/130117-01. January.

Commission on the Prevention of WMD Proliferation and Terrorism. 2008. *World At Risk*. http://www.pharmathene.com/World_at_Risk_Report.pdf.

Cullen, Lord W. Douglas. 1990. *The Public Inquiry into the Piper Alpha Disaster*. 2 vol. London: HM Stationery Office.

Engineering Failures. 2009. Engineering Failures: Chernobyl. Case Studies in Engineering. http://engineeringfailures.org/?p=1 (posted July 10, 2009).

European Committee for Standardization. 2011. CEN Workshop Agreement (CWA) 15793— Laboratory Biorisk Management.

Federal Bureau of Investigation. 2011. Amerithrax or Anthrax Investigation. http://www.fbi.gov/anthrax/amerithraxlinks.htm (accessed August 12, 2014).

Fonkwo, Peter. 2008. Biosecurity Challenges of the Global Expansion of High-Containment Biological Laboratories. *EMBO Reports*, 9(Suppl 1): S13–S17. doi: 10.1038/embor. 2008.110. http://www.ncbi.nlm.nih.gov/pmc/articles/PMC3327542/.

Friedman, M.A., J. Woodcock, M.M. Lumpkin, J.E. Shuren, A.E. Hass, and L.J. Thomson. 1999. The Safety of Newly Approved Medicines: Do Recent Market Removals Mean There Is a Problem? *Journal of the American Medical Association*, 12(281): 18.

Furmanski, Martin. 2014. Laboratory Escapes 'Self-Fulfilling Prophecy' Epidemics. Center for Arms Control and Nonproliferation. February 17. http://armscontrolcenter.org/Escaped_Viruses-final_2-17-14.pdf.

Global Food Safety Initiative. 2014. http://www.mygfsi.com (accessed April 2014).

Government of Japan. 2007. Amendment of the Infectious Diseases Control Law, Japan, as of June 2007. *Infectious Agents Surveillance Report*, 28: 185–188. http://idsc.nih.go.jp/iasr/28/329/tpc329.html.

Government of South Korea. 2005. Prevention of Contagious Disease Act.

Graham, Nancy. 2010. Aviation Safety: Making a Safe System Even Safer. Video message. International Civil Aviation Organization. October 1. http://www.icao.int/Newsroom/Presentation%20Slides/Streaming%20video%20message%20-%20Aviation%20Safety.pdf.

Harris, Anne. 2013. Piper Alpha 25 Years On—Have We Learned the Lessons? *Engineering and Technology Magazine*, 8: 7. http://eandt.theiet.org/magazine/2013/07/never-stop-learning.cfm.

Health and Safety Executive. 2007. Final Report on Potential Breaches of Biosecurity at the Pirbright Site 2007. December 20. http://www.hse.gov.uk/news/2007/finalreport.pdf (accessed August 12, 2014).

Henkel, Richard D., T. Miller, and R.S. Weyant. 2012. Monitoring Select Agent Theft, Loss and Release Reports in the United States—2004–2010. *Applied Biosafety*, 17(4).

Herfst, Sander, E.J.A. Schrauwen, M. Linster, S. Chutinimitkul, E. de Wit, V.J. Munster, E.M. Sorrell, T.M. Bestebroer, D.F. Burke, D.K. Smith, G.F. Rimmelzwaan, A.D.M.E. Osterhaus, and R.A.M. Fouchier. 2012. Airborne Transmission of Influenza A/H5N1 Virus between Ferrets. *Science*, 336(6088): 1534–1541.

ICH (International Conference on the Harmonisation of Technical Requirements for Registration of Pharmaceuticals for Human Use). 2014. History/About ICH. http://www.ich.org/about/history.html (accessed September 20, 2014).

Imai, Masaki, T. Watanabe, M. Hatta, S.C. Das, M. Ozawa, K. Shinya, G. Zhong, A. Hanson, H. Katsura, S. Watanabe, C. Li, E. Kawakami, S. Yamada, M. Kiso, Y. Suzuki, E. Maher, G. Nermann, and Y. Kawaoka. 2012. Experimental Adaptation of an Influenza H5 HA Confers Respiratory Droplet Transmission to a Reassortant H5 HA/H1N1 Virus in Ferrets. *Nature*, 486(7403): 420–428.

Institute of Medicine. 2006. *The Future of Drug Safety: Promoting and Protecting the Health of the Public*. Washington, DC: National Academies Press. http://www.iom.edu/Reports/2006/The-Future-of-Drug-Safety-Promoting-and-Protecting-the-Health-of-the-Public.aspx.

International Atomic Energy Agency. 2006. Chernobyl's Legacy: Health, Environmental, and Socio-Economic Impacts. In *The Chernobyl Forum: 2003–2005*. 2nd rev. version. http://www.iaea.org/Publications/Booklets/Chernobyl/chernobyl.pdf.

International Civil Aviation Organization. 2007. Global Aviation Safety Plan. July. http://www.icao.int/WACAF/AFIRAN08_Doc/gasp_en.pdf.

International Civil Aviation Organization. 2013. *Safety Management Manual*. Document 9859, AN/474, 3rd ed. http://www.skybrary.aero/bookshelf/content/bookDetails.php?bookId=644.

International Nuclear Safety Advisory Group. 1992. INSAG-7: The Chernobyl Accident—Updating of INSAG-1. Safety Series 75. International Atomic Energy Agency, Vienna. http://www-pub.iaea.org/MTCD/publications/PDF/Pub913e_web.pdf.

International Organization for Standardization. 2005. ISO 22000: Food Safety Management. http://www.iso.org/iso/home/standards/management-standards/iso22000.htm (accessed April 2014).

Karasek, Gary. 2014. Top 10 Major Accidents That Influenced the World. Parts 3–5. http://www.actiononrisk.com/2011/01/top-10-major-accidents-that-influenced-the-world-part-3/, http://www.actiononrisk.com/2011/01/top-10-major-accidents-that-influenced-the-world-part-4/, http://www.actiononrisk.com/2011/04/top-10-major-accidents-that-influenced-the-world-part-5/ (accessed April 2014).

Kingdom of Denmark. 2008. Act on Securing Certain Biological Agents, Delivery Systems and Related Material. Law 69. Adopted by the Danish Parliament on the 3rd hearing, June 12.

Kortepeter, Mark G., J.W. Martin, J.M. Rusnak, T.J. Cieslak, K.L. Warfield, E.L. Anderson, and M.V. Ranadive. 2008. Managing Potential Laboratory Exposure to Ebola Virus by Using a Patient Biocontainment Care Unit. *Emerging Infectious Diseases*, 14(6). doi: 10.3201/eid1406.071489.

Kruse, Richard H., W.H. Puckett, and J.H. Richardson. 1991. Biological Safety Cabinet. *Clinical Microbiology Reviews*, 207–241.

Lenz, W. 1988. A Short History of Thalidomide Embryopathy. *Teratology*, 38: 3.

Lind, Arne. 1957. Ventilated Cabinets in a Tuberculosis Laboratory. *Bulletin of the World Health Organization*, 16(2): 448–453.

Lipsitch, Marc, and A. Galvani. 2014. Ethical Alternatives to Experiments with Novel Potential Pandemic Pathogens. *PLoS Medicine*, 10.1371.

Manaan, M. Sam, ed. 2005. *Lee's Loss Prevention in the Process Industries: Hazard Identification, Assessment, and Control*. 3rd ed., Vol. 3, Appendix 5—Bhopal.

Manaan, M. Sam, H.H. West, K. Krishna, A.A. Aldeeb, N. Keren, S.R. Saraf, Y.S. Liu, and M. Gentile. 2005. The Legacy of Bhopal: The Impact over the Last 20 Years and Future Direction. *Journal of Loss Prevention in the Process Industries*, 18: 218–224.

Manning, Shannon. 2010. *Escherichia coli Infections*. 2nd ed. New York: Chelsea House.

Margolin, Josh, and T. Sherman. 2005. 3 Plague-Infected Lab Mice Missing. *Seattle Times*. http://seattletimes.com/html/nationworld/2002498338_plague16.html (posted September 16).

McGreal, Ryan. 2012. 17 Months Later, Fukushima Daiichi Offers Bitter Lesson in Risk Management. *Raise the Hammer*. August 9. https://raisethehammer.org/article/1641/17_months_later_fukushima_daiichi_offers_bitter_lessons_in_risk_management.

Meadows, Michelle. 2006a. A Century of Ensuring Safe Foods and Cosmetics. *FDA Consumer Magazine*, The Centennial Edition. January-February. http://www.fda.gov/AboutFDA/WhatWeDo/History/FOrgsHistory/CFSAN/ucm083863.htm.

Meadows, Michelle. 2006b. Promoting Safe and Effective Drugs for 100 Years. *FDA Consumer Magazine*. January-February. http://www.fda.gov/AboutFDA/WhatWeDo/History/ProductRegulation/PromotingSafeandEffectiveDrugsfor100Years/.

Mermel, Leonard A., S.L. Josephson, J. Dempsy, S. Parenteau, C. Perry, and N. Magill. 1997. Outbreak of *Shigella sonnei* in a Clinical Microbiology Laboratory. *Journal of Clinical Microbiology*, 35(12), 3163–3165.

Miller, Judith. 2004. Russian Scientist Dies in Ebola Accident at Former Weapons Lab. *New York Times*. May 25. http://www.nytimes.com/2004/05/25/international/europe/25ebol.html (accessed August 12, 2014).

Miller, Marilyn T., and K. Strömland. 1999. Thalidomide: A Review, with a Focus on Ocular Findings and New Potential Uses. *Teratology*, 60: 3.

Mouawad, Jad, and C. Drew. 2013. Airline Industry at Its Safest since the Dawn of the Jet Age. *New York Times*. February 11. http://www.nytimes.com/2013/02/12/business/2012-was-the-safest-year-for-airlines-globally-since-1945.html?pagewanted=all&_r=0.

National Aeronautics and Space Administration. 2013. *The Case for Safety: The North Sea Piper Alpha Disaster.* NASA Safety Center System Failure Case Study 7: 4. May.

National Diet of Japan. 2012. The Official Report of the Fukushima Nuclear Accident Independent Investigation Commission. http://warp.da.ndl.go.jp/info:ndljp/pid/3856371/naiic.go.jp/wp-content/uploads/2012/09/NAIIC_report_lo_res10.pdf.

National Research Council. 2004. Biotechnology Research in the Age of Bioterrorism. Committee on Research Standards and Practices to Prevent the Destructive Application of Biotechnology, Washington, DC.

National Research Council. 2010. Sequence-Based Classification of Select Agents: A Brighter Line. Committee on Scientific Milestones for the Development of a Gene-Sequence-Based Classification System for the Oversight of Select Agents, Washington, DC.

Oil and Gas Industry Association. 2013. Piper 25, Presented at Oil and Gas UK Annual Conference, Aberdeen, UK, June 18–20. http://www.oilandgasuk.co.uk/events/Piper25.cfm?frmAlias=/Piper25/ (accessed April 2014).

O'Mahony, Ann. 2011. Quality Risk Management: The Pharmaceutical Experience. PDA presentation in Galway, Ireland. November 11. http://www.pda.org/docs/default-source/website-document-library/chapters/presentations/ireland/quality-risk-management—-the-pharmaceutical-experience.pdf?sfvrsn=6.

Osterholm, Michael T. 2011. Foodborne Disease in 2011—The Rest of the Story. *New England Journal of Medicine*, 364: 10.

Palk, Justin M. 2009. USAMRIID Finds 9,200 Disease Samples It Didn't Know It Had. *Frederick News-Post*. http://www.fredericknewspost.com/archive/article_dd4e2f57-6720-5615-9a76-bdf5c443c333.html?mode=jqm (posted June 18).

Pauwels, Katia, R. Mampuys, C. Golstein, D. Breyer, P. Herman, M. Kaspari, J.C. Pagés, H. Pfister, F. van der Wilk, and B. Schönig. 2013. Event Report: SynBio Workshop (Paris 2012)—Risk Assessment Challenges of Synthetic Biology. *Journal of Consumer Protection and Food Safety*, 8(3): 215–226.

Phillips, G. Brooks. 1961. Technical Study 35: Microbiological Safety in U.S. and Foreign Laboratories. Fort Detrick, MD.

Pike, Robert M. 1976. Laboratory-Acquired Infections: Summary and Analysis of 3,921 Cases. *Health Laboratory Science*, 13(2): 105–114.

Public Health Agency of Canada. 2014. Compliance: Registration, Permits, Inspection and Enforcement. http://www.phac-aspc.gc.ca/lab-bio/permits/index-eng.php (last modified July 17, 2014).

Reitman, Morton, and A.G. Wedum. 1966. Microbiological Safety. *Public Health Report*, 71: 659–665.

Republic of Singapore. 2005. Biological Agents and Toxins Act. Bill 26/2005. Passed September 20. http://www.biosafety.moh.gov.sg/bioe/ui/pages/links/abt_bata.htm.

Reuters. 2012. Utility Says It Underestimated Radiation Released in Japan. *New York Times*. May 24. http://www.nytimes.com/2012/05/25/world/asia/radioactive-release-at-fukushima-plant-was-underestimated.html?_r=0.

Russ, Hillary, and J. Steenhuysen. 2014. CDC Reassigns Director of Lab Behind Anthrax Blunder. Reuters. June 24. http://www.reuters.com/article/2014/06/23/us-usa-anthrax-idUSKBN0EY0A020140623.

Saxholm, Rolf. 1954. Experiments with a New Culture Method for Tubercle Bacilli. *American Review of Tuberculosis*, 69: 304–306.

Scallan, Elaine, R.M. Hoekstra, F.J. Angulo, R.V. Tauxe, M.A. Widdowson, S.L. Roy, J.L. Jones, and P.M. Griffin. 2011. Foodborne Illness Acquired in the United States—Major Pathogens. *Emerging Infectious Diseases*, 17: 1. doi: 10.3201/eid1701.091101p1.

Scott, Willie. 2011. Piper Oil Rig Disaster. Bright Hub Engineering. http://www.brighthubengineering.com/marine-history/116049-piper-alpha-oil-rig-disaster/ (updated November 10, 2011).

Sherman, Ted. 2009. UMDNJ Facility Loses Two Plague-Infected Dead Lab Mice. *The Star-Ledger*. http://www.nj.com/news/index.ssf/2009/02/dead_lab_mice_lost_from_umdnj.html (posted February 7, 2009).

Shooter, Reginald A. 1980. *Report of the Investigation into the Cause of the 1978 Birmingham Smallpox Occurrence*. London: HMSO.

Srinivansan, Arjun, C.N. Kraus, D. DeShazer, P.M. Becker, J.D. Dick, L. Spacek, J.G. Bartlett, W.R. Byrne, and D.L. Thomas. 2001. Glanders in a Military Research Microbiologist. *New England Journal of Medicine*, 345(4): 256–258.

Starr, Steven. 2012. Costs and Consequences of the Fukushima Daiichi Disaster. Physicians for Social Responsibility. http://www.psr.org/environment-and-health/environmental-health-policy-institute/responses/costs-and-consequences-of-fukushima.html (posted November 2012).

Steenhuysen, Julie, and S. Begley. 2014. CDC Didn't Heed Own Lessons from 2004 Anthrax Scare. Reuters. June 30. http://www.reuters.com/article/2014/06/29/us-usa-anthrax-risks-insight-idUSKBN0F40DY20140629.

The Sunshine Project. 2007. News Release: Texas A&M Bioweapons Accidents More the Norm than the Exception. July 3.

Tanne, Janice H. 2003. Infectious Diseases Expert Convicted over Missing Plague Bacteria. *BMJ*. doj: 10.1136/bmj.327.7427.1307-b. http://www.ncbi.nlm.nih.gov/pmc/articles/PMC1146509/.

ten Berge, Dieudonne. 1990. *The First 24 Hours: A Comprehensive Guide to Successful Crisis Management*. Colchester, VT: Blackwell Business.

Ten Hoeve, John E., and M.Z. Jacobson. 2012. Worldwide Health Effects of the Fukushima Daiichi Nuclear Accident. *Energy and Environmental Science*. June. doi: 10.1039/c2ee22019a. http://web.stanford.edu/group/efmh/jacobson/TenHoeveEES12.pdf.

University of Edinburgh. 2009. Variant Creutzfeld-Jakob Disease: Current Data (October 2009). National Creutzfeldt-Jakob Disease Surveillance Unit (NCJDSU), University of Edinburgh. October.

US Centers for Disease Control and Prevention. 1974. *Classification of Etiological Agents on the Basis of Hazard*. 4th ed.

US Centers for Disease Control and Prevention. 1993. Update: Multistate Outbreak of *Escherichia coli* O157:H7 Infections from Hamburgers—Western United States, 1992–1993. *Morbidity and Mortality Weekly Report*, 42(14).

US Centers for Disease Control and Prevention. 2004. China Reports Ninth Recent Possible SARS Case. Health Advisory. April 29.

US Centers for Disease Control and Prevention, DSAT. 2007. Letter from DSAT Director to Responsible Official, Texas A&M University. Division of Select Agents and Toxins. August 31.

US Chemical Safety and Hazard Investigation Board. 2007. *Investigation Report: Refinery Explosion and Fire, Texas City, Texas, March 23, 2005*. Report 2005-04-I-TX. March. http://www.csb.gov/assets/1/19/csbfinalreportbp.pdf.

US Code of Federal Regulations. 1996. 42 CFR Part 72: Additional Requirements for Facilities Transferring or Receiving Select Agents. October 24, 1996.

US Code of Federal Regulations. 2005. 42 CFR Part 73: CDC Select Agent Regulations. 7 CFR Part 331 and 9 CFR Part 121: APHIS Select Agent Regulations. http://www.selectagents.gov/Regulations.html.

US Department of Army. 2014. The History of Fort Detrick. http://www.detrick.army.mil/cutting_edge (accessed April 11, 2014).

US Department of Energy. 2012. *Special Report: Inquiry into the Security Breach at the National Nuclear Security Administration's Y-12 National Security Complex.* DOE/IG-0868. Office of the Inspector General. August. http://energy.gov/sites/prod/files/IG-0868_0.pdf.

US Department of Health and Human Services. 1984. *Biosafety in Microbiological and Biomedical Laboratories.* 1st ed.

US Department of Health and Human Services. 2009. *Biosafety in Microbiological and Biomedical Laboratories.* 5th ed., p. 3.

US Federal Aviation Administration. 2010. Press Release—FAA Proposes Safety Management Systems for Airlines. November 4. https://www.faa.gov/news/press_releases/news_story.cfm?newsid=12118.

US Food and Drug Administration. 2014. Snapshot of Food Safety Milestones in the History of the FDA. http://www.fda.gov/Food/GuidanceRegulation/FSMA/ucm238505.htm (last updated August 5).

US National Institutes of Health. 1976. Recombinant DNA Research Guidelines. *Federal Register,* 41(131): 27902–27943.

US Senate. 2006. Food Safety: Current Challenges and New Ideas to Safeguard Consumers. Hearing of the Committee on Health, Education, Labor, and Pensions, US Senate. November 15. http://www.gpo.gov/fdsys/pkg/CHRG-109shrg31620/html/CHRG-109shrg31620.htm.

US White House. 2010. Executive Order 13546—Optimizing the Security of Biological Select Agents and Toxins in the United States. http://www.whitehouse.gov/the-press-office/executive-order-optimizing-security-biological-select-agents-and-toxins-united-stat.

Valleron, Alain-Jacques, P.Y. Boelle, R. Will, and J.Y. Cesbron. 2001. Estimation of Epidemic Size and Incubation Time Based on Age Characteristics of vCJD in the United Kingdom. *Science* 294(5547). doi: 10.1126/science.1066838.

Van Noorden, Richard. 2013. Safety Survey Reveals Lab Risks. *Nature,* 493: 3.

Walker, J. Samuel, and T.R. Wellock. 2010. *A Short History of Nuclear Regulation, 1946–2009.* US Nuclear Regulatory Commission. October. http://pbadupws.nrc.gov/docs/ML1029/ML102980443.pdf.

Watanabe, Tokiko, G. Zhong, C.A. Russell, N. Nakajima, M. Hatta, A. Hanson, R. McBride, D.F. Burke, K. Takahashi, S. Fukuyama, Y. Tomita, E.A. Maher, S. Watanabe, M. Imai, G. Neumann, H. Hasegawa, J.C. Paulson, D.J. Smith, and Y. Kawaoka. 2014. Circulating Avian Influenza Viruses Closely Related to the 1918 Virus Have Pandemic Potential. *Cell Host and Microbe,* 15(6): 692–705.

Wedum, Arnold G. 1997. History and Epidemiology of Laboratory-Acquired Infections. *Journal of the American Biological Safety Association,* 2(1): 12–29.

Wikipedia. 2014. Texas City Refinery Explosion. http://en.wikipedia.org/wiki/Texas_City_Refinery_explosion (accessed April 2014).

World Health Organization. 1983. *Laboratory Biosafety Manual.* 1st ed.

World Health Organization. 2004. *Laboratory Biosafety Manual.* 3rd ed.

World Health Organization. 2003. *WHO Good Manufacturing Practices: Main Principles for Pharmaceutical Products.* 37th Report. WHO Expert Committee on Specifications for Pharmaceutical Preparations, Geneva, Annex 4.

2 The AMP Model

Lisa Astuto Gribble, Edith Sangalang Tria, and Laurie Wallis

ABSTRACT

It is the responsibility of all organizations that work with biological agents and toxins to operate safely and securely. Biorisk management, as defined by CEN Workshop Agreement (CWA) 15793:2011, is "a system or process to control safety and security risks associated with the handling or storage and disposal of biological agents and toxins in laboratories and facilities." Effectively implementing this type of management is a complex process that involves all organizational stakeholders and institutional levels. It takes time, resources, and continual oversight and effort to create and sustain a highly effective biorisk management system in any organization.

The AMP model is a simple yet effective method for supporting the implementation of biorisk management. The model is composed of three basic components: assessment (A), mitigation (M), and performance (P). No biorisk management system is complete or comprehensive without the inclusion of these three components.[*]

INTRODUCTION

Organizations that work with biological agents and toxins have a responsibility to operate safely and securely. Achieving safety and security requires the management of all biological risks, whether they are in a laboratory, hospital, or occupational health setting. The CEN Workshop Agreement (CWA) 15793:2011 defines biorisk management as "a system or process to control safety and security risks associated with the handling or storage and disposal of biological agents and toxins in laboratories and facilities" (European Committee for Standardization 2011).

Biorisk management is a new field; its formal origin dates back to the early 2000s, following a number of dangerous biological laboratory incidents, such as the Amerithrax attacks of 2001 and laboratory-acquired severe acute respiratory syndrome (SARS) infections in Asia during 2003–2004. In response, a concerned international scientific and policy community sought to create a harmonized biological risk management approach to increase awareness of biological risks and to establish improved conformity of biosafety and biosecurity activities around the world.

Biorisk management can be divided into three primary components: assessment (A), mitigation (M), and performance (P). None is a new concept. In fact, each has been adopted independently by various industrial sectors for decades. The

---

[*] The AMP model was first articulated by the World Health Organization in its Biorisk Management Advanced Trainer Programme, developed and first executed in 2010.

identification and *assessment* of risk has a long history, but it was not formally recognized until the early 1980s (Environmental Protection Agency 2004; Kaplan and
Garrick 1981). Since this time, the field of risk analysis and assessment has expanded
and has become an integral part of numerous businesses and industries. As discussed in the introduction, the airline industry is a good example of an industry
that relies heavily on risk assessment, since it is exposed to substantial risk in its
day-to-day operations, as well as risks associated with customer safety and relations,
corporate reputation and value, and aircraft maintenance and security, among others.
Historically, the airline industry's safety was built upon a reactive analysis of past
catastrophic aircraft accidents. Today, with the introduction of risk analysis and the
identification and management of all areas of risk, the international airline industry has achieved a remarkably high level of safety (International Civil Aviation
Organization 2013).

Mitigation strategies represent perhaps the most common management approach
to achieve safety and security. The creation, design, development, and sale of risk control measures encompass a billion-dollar enterprise that spreads across many industries. Mitigation control measures offer indispensable tools and practices to reduce
risks. Historically, businesses and industries have first sought mitigation measures
that will directly reduce or eliminate their risks. Certainly, it is easier to purchase a
specific technology or device to reduce risk than to strategize about how an organization with multiple stakeholders thinks about risk; the latter approach often requires a
paradigm shift in how a company manages and meaningfully reduces its risks.

Lastly, it is critical for an organization's overall mission and objectives to understand its safety record and how to best evaluate its safety *performance*. Even though
many industries rely on performance metrics, such as tracking "days without a
safety incident," less than assessment and mitigation processes, the healthcare industry has relied on evaluating its performance and the delivery of quality care for
nearly 250 years (Loeb 2004).

Clearly, assessment, mitigation, and performance represent critical elements in
any business process, including risk management. However, in a biorisk management system, each of the three components is not addressed individually, but is collectively captured by what is called the AMP model (World Health Organization
2010) shown in Figure 2.1. The AMP model requires that control measures be based
on a robust risk assessment, and the effectiveness and suitability of the control measures be continually evaluated. Identified risk can be either mitigated (avoided, limited, or transferred to an outside entity) or accepted.

Each component of the AMP model contributes equally to an effective biorisk
management system. Like a three-legged stool, a biorisk management system fails
if one of the components, or legs, is overlooked or is not addressed in a meaningful
and comprehensive way. In contrast to other risk management models, which have
typically focused heavily on mitigation measures, AMP focuses equal attention on
assessment and performance. Although simple, the AMP model is vital whenever
implementing biorisk management. After completing the initial round of the assessment, mitigation, and performance steps—establishing a snapshot in time—the
biorisk management system continually reassesses and reevaluates the current system and make changes, as needed.

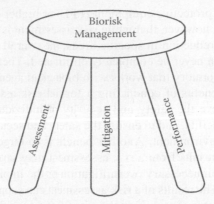

Biorisk
Management

Assessment

Mitigation

Performance

FIGURE 2.1 The AMP model. Without each leg supporting biorisk management, the stool will fail.

The following sections of this chapter will introduce and describe each of the three AMP components. More detailed explanations are provided in the subsequent chapters.

ASSESSMENT

The first fundamental component of a biorisk management system begins with an assessment of the risks present in the laboratory. Risk can be described as the combination of the likelihood (or probability) and the consequences of an undesirable event. It is often described as a mathematical equation: risk $= f$ (likelihood, consequences). A risk can be based on either a hazard or a threat.[*][†] A risk assessment is the fundamental process to help determine, mitigate, and manage laboratory risks. A good risk assessment system informs decisions intended to reduce the risks present in a laboratory. Chapter 3, "Risk Assessment," provides a more detailed review of the risk assessment methodology.

The results of a risk assessment guide the selection of appropriate biological safety measures (including microbiological practices and selection of proper safety equipment), security measures (including controlled access to the laboratory where the biological agents exist), and other facility safeguards to mitigate risks to an acceptable or manageable level. In many instances, the results of a well-executed risk assessment will demonstrate that some risks can be controlled using relatively straightforward measures, such as properly cleaning up spills and splashes, reducing fall hazards, and locking storage areas that contain infectious pathogens. Controlling other risks, of course, may require a greater investment of resources to mitigate them appropriately. In parallel, risk assessment results should shape the objectives of laboratory training. If the results of a risk assessment show a low likelihood and consequence of infection, assuming the use of minimal personal protective equipment such as gloves and gowns, staff will obviously not need to be trained in the use

---

[*] A hazard is a source, situation, or act with the potential for harm.
[†] A threat is a person who has intent or ability to cause harm.

of expensive personal protective equipment (PPE) or higher containment practices. It is important to note, however, that while risk assessment is an essential tool to aid in risk elimination or reduction in specific areas, the overall laboratory risk, by the nature of the work, can never be completely eliminated. There will always be some level of risk in any laboratory that works with biological agents and toxins.

There are many benefits of conducting a formal risk assessment. The primary advantage is to improve the safety and security of individuals and the biological materials within the facility, and to ensure the safety and security of the surrounding community and the environment. Another benefit is to target mitigation resources most effectively. The results from a risk assessment may save the company money, preventing costly and unnecessary overmitigation or the implementation of improperly applied controls. The results of a risk assessment can support advanced planning of renovations to laboratory space, including justifying specific facility or equipment needs. The risk assessment process can also be used to evaluate and validate emergency plans, as well as plan for preventive maintenance of critical equipment and facility features.

The quality of a risk assessment's results is entirely dependent upon the quality of the information collected while conducting the risk assessment. In other words, a risk assessment requires the collection and evaluation of accurate information. Personnel assigned to contribute to a risk assessment should be intimately familiar with the laboratory's work activities, its biological agent holdings, procedures, equipment, and personnel, and how they all affect the laboratory's risks. All information that feeds into the risk assessment process must be collected and assessed by those in the facility who are involved in managing biorisks, including laboratory managers, principal investigators, laboratory staff, and safety and security professionals, among others.

Many in the international community recognize the importance of risk assessment to reduce biological laboratory risks. Biological risk assessment is a legal obligation in many countries that have biosafety regulations, as part of the notification or authorization process or as a basis to determine the required risk mitigation measures. Many leading guidance documents on biosafety, including the World Health Organization's (WHO) *Laboratory Biosafety Manual*, the US *Biosafety in Microbiological and Biomedical Laboratories*, and the international consensus document "CEN Workshop Agreement (CWA) 15793:2011—Laboratory Biorisk Management" (referred to as CWA 15793), emphasize that risk assessment is the fundamental planning step for managing these risks (European Committee for Standardization 2011; World Health Organization 2004; US Department of Health and Human Services 2009).

It is important to note that a robust risk assessment must be specific to the unique laboratory, situation, or facility. Every facility has different features and equipment, and every institution has a different perspective on risk mitigation or acceptance. The procedures used in the facilities, and the knowledge, skills, and abilities of the individuals performing them, are almost never all identical. A generic agent risk assessment (e.g., pathogen safety data sheets distributed by the Public Health Agency of Canada (Public Health Agency of Canada 2014)) alone is not sufficient for biorisk management because it evaluates only the agent—not any of the other factors that impact the risk. However, these descriptions of the agent characteristics often provide background information that can enhance a risk assessment.

WHEN TO PERFORM AND REVIEW A LABORATORY RISK ASSESSMENT

When experiments, processes, materials, and technology change, so does the risk. A laboratory should always perform and document a risk assessment before initiating any new work. A risk assessment should also be completed whenever a noteworthy change occurs at the facility or there is a change in the basic nature of the work being conducted. Regardless, even in static conditions, it is important to conduct a periodic assessment of laboratory risk. Biological risks can still change even if the experiment or processes do not; therefore, a risk assessment should be performed and reviewed at least annually. Examples of activities or events that will change the risk environment and warrant a reassessment include:

- New infectious agents, toxins, reagents, or other dangerous substances
- New animal species, model, or route of administration of biological agents
- Different procedures and practices
- New equipment
- Changes in personnel
- Changes in manufacturer or supplier of consumable materials (PPE, containers, waste disposal materials, media, etc.)
- Equipment that may no longer be operating effectively because it has deteriorated or has not had adequate repair/maintenance
- Advances in scientific understanding and technology (new paradigms)
- A relocation or renovation
- A recent accident, laboratory-acquired infection (LAI), theft, or security violation
- National or regional changes in disease status (endemicity of disease or disease eradication)
- Changes in reliable local infrastructure (electricity, water, roads)
- National, regional, or local changes in the threat or security environment

After reviewing the results of the risk assessment, appropriate mitigation measures should be identified, implemented, or amended, as necessary.

SHARED ROLES AND RESPONSIBILITIES IN ASSESSING RISK

The risk assessment process should not be driven or executed solely by a laboratory's biorisk management advisor or biosafety professional; rather, *a quality risk assessment is the culmination of input from numerous people in the laboratory or facility.*

Further, *risk assessment is a critical responsibility* for individuals within the laboratory and the larger facility. In an individual laboratory, the best assessors of risk are usually those who work in the laboratory and who are most familiar with the agents and other valuable laboratory materials, as well as the experimental practices and processes. Thus, laboratory biosafety and biosecurity risk assessments should be a shared responsibility between principal investigators, scientists, researchers (or a risk assessment team), and the biorisk management advisors or biosafety professionals. Biosafety professionals should assume responsibility for initiating the risk assessment process and remain highly vigilant regarding their awareness of all biorisks present within the institution's laboratories. For a biosecurity risk assessment, institution security professionals should also be involved, whenever possible. Descriptions of the various risk assessment users and their responsibilities appear in Chapter 3.

Once all of the risks have been identified and communicated, the relevant laboratory and institutional staff should work together toward controlling or reducing the risks efficiently to an acceptable level. Laboratory staff and other stakeholders can work together to determine the necessary biosafety practices required to perform a task with a specific organism. This should be done in close collaboration with other responsible institutional parties, such as an institutional biosafety committee (IBC) or biorisk management committee, a hospital's infection control committee and patient's safety committee, a department of environmental health and safety, a laboratory quality assurance committee, or any animal care and use committees.

MITIGATION

The second fundamental component of the biorisk management model is mitigation. Biorisk mitigation measures are actions and control measures, based on a robust laboratory risk assessment, that are put into place to reduce or eliminate the risks associated with biological agents and toxins. Assessing the risks determines the actions and control measures that will be most effective in reducing and eliminating those particular risks. Often facilities implement unnecessary or inappropriate mitigation measures because a risk assessment has not yet been completed or was completed poorly. In many of those cases, such measures do not reduce the risks, but can rather increase them. An example includes the common mistake of a laboratory that is persuaded by a commercial vendor into purchasing an unnecessary and expensive Class III glove box, but it fails to adequately train staff on how to use the cabinet or conduct routine maintenance or certification. This situation could leave the laboratory staff more prone to infection than before the equipment was purchased.

Historically, the traditional biosafety approach has emphasized physical mitigation measures while ignoring, or not properly identifying, equally important factors. For example, many facilities have invested significant resources to purchase and install video cameras to reduce the biosafety and biosecurity risks present in their laboratories. It is common for bioscience facilities to purchase sophisticated equipment or invest in a training program for new employees. The types of mitigation measures available today are numerous and varied, and indeed, many play a vital role in reducing biorisks. However, it is critical to realize that *mitigation measures alone are not sufficient for an effective biorisk management program*. Mitigation is just one component of the biorisk management process.

Mitigation Control Measures

Overall, biorisk mitigation can be divided into five areas of control. The first category of mitigation control is elimination or substitution. Elimination involves not doing the intended work, or deciding not to work with a specific biological agent. Obviously, elimination provides the highest degree of risk reduction. A well-known global example of elimination is the international community's decision to consolidate and retain all known stocks of Variola major, the biological agent that causes smallpox, in only two WHO reference laboratories in the world—the US Centers for Disease Control and Prevention (CDC) in Atlanta, Georgia, and the State Research Center of Virology and Biotechnology (VECTOR) in Russia. This decision eliminated the risk involved with smallpox research at all but two laboratories in the world. However, in many situations, eliminating the risk is not always feasible. For those cases, it may be necessary to use a substitute, or to replace or exchange the source of the identified risk with another source that poses less of a hazard/threat than the original risk. For example, a laboratory conducting research with the pathogen *Bacillus anthracis*, responsible for causing the acute fatal disease anthrax, could potentially substitute a less dangerous experimental surrogate, such as *Bacillus thuringiensis*, an organism most commonly used in biological pesticides worldwide. This decision would significantly reduce the risk of infection while perhaps not compromising the research objectives. In addition, the use of *B. thuringiensis* also eliminates the laboratory's need for elevated biosafety and containment handling.

The second category of mitigation control measures is engineering controls. These control measures are physical changes to work stations, equipment, production facilities, or any other relevant aspect of the work environment that reduces or prevents exposure to hazards. A biosafety cabinet, which comes in three levels of protection, is an example of an engineering control; Class I and II cabinets are designed with unidirectional, laminar airflow to direct potentially contaminated air away from workers and through HEPA filters before exiting to the environment. The Class III cabinets add additional rigorous containment, using gas-tight glove boxes and other features. Even the simple method of locking laboratory doors is an example of security-related engineering controls.

The third category of mitigation control is collectively called administrative controls. These controls are policies, standards, and guidelines used to control risks. Proficiency and competency training for laboratory staff would be considered an administrative control. Displaying biohazard or warning signage, markings, and labeling, controlling visitor and worker access, and documenting written standard operating procedures are all forms of administrative controls.

Practices and procedures comprise the fourth category of mitigation control measures. This includes practices to minimize splashes, sprays, and aerosols to avoid laboratory-acquired infections or following standard operating procedures (SOPs), for example.

The final group of mitigation control measures is personal protective equipment (PPE). These are devices worn by workers to protect them against chemicals, toxins, and pathogenic hazards in the laboratory. Gloves, gowns, and respirators are all examples of PPE. PPE is considered the least effective control because it only protects the person who is wearing it, and only if it is used correctly. Its failure or inappropriate use, a rip in the material, or a manufacturing defect, for example, would likely result in exposure.

Since risk is a function of likelihood and consequences, it is useful to recognize how mitigation measures impact the equation of risk. The implementation of engineering controls, administrative controls, practices and procedures, and PPE should decrease the likelihood of risk. Substitution will affect the consequence side of the equation; elimination of the hazard will eliminate the risk altogether. Assessing the risk and understanding the likelihood and consequences of various risks are important because they establish the basis for strategic decisions about control measures.

While all five types of mitigation control measures are important and can contribute to reducing biorisks, not one is completely effective at controlling or reducing *all* risks. Moreover, there are distinct advantages and disadvantages to each approach. For instance, an air handling system for the lab rooms may effectively reduce the risk that an airborne pathogen could escape from the laboratory, but that system may do little to protect the employees inside the laboratory. Also, that system may be expensive to install, operate, and maintain, and require a consistent, dedicated power supply. For more detailed examples of mitigation measures, as well as their advantages and disadvantages, see Chapter 5, "Rethinking Mitigation Measures."

In general, the most effective way to mitigate risk is to consider a combination of controls. The concept of a hierarchy of controls describes an order of effectiveness (from most effective to least effective) for mitigation measures, and implies that this order should be taken into account when selecting and implementing controls to reduce risk. However, depending on the facility or situation, a mitigation measure lower in the hierarchy may actually be more effective than one that is higher in the hierarchy. Decision makers must also assess organizational strengths, resources, commitment, personnel knowledge, and competency, among other attributes, to ensure that the mitigation measures are implemented appropriately for the application.

Clearly, many facilities and laboratories around the world face challenges in mitigating biorisks because they have minimal organizational and financial resources. Many laboratories do not have the institutional guidance to address their safety or security risks, or the programs and management systems they have in place are *ad hoc* and not fully supported by management. Other difficulties that confront laboratories around the world include the absence of a sustainable electrical supply, inadequate facility infrastructure, security concerns related to geographic areas, volatile weather conditions and geologic hazards, inadequately trained personnel, and ambiguous or nonexistent national and international regulations or guidance. Nonetheless, when laboratory personnel decide to utilize a risk-based approach to mitigate the biorisks present in their own laboratories, they become empowered to better understand the safety and security risks that directly affect workers, as well as their families and communities. A risk-based approach can also enable personnel to take the necessary measures to reduce those risks in a manner that makes the most sense in their specific environment, rather than being held to a prescriptive risk mitigation approach, which may not be attainable or sustainable in all labs, whether in developed or developing regions. Thus, it is important to recognize that specific risk reduction measures will vary significantly from laboratory to laboratory, institution to institution, country to country, and region to region.

PERFORMANCE

Performance represents the third pillar of the biorisk management model. Performance management is a systematic process intended to achieve improved levels of organizational objectives and goals. An institution's ability to manage and evaluate its performance contributes directly to its development and improvement. Management systems with this insight have a direct advantage over organizations that do not. A number of studies have clearly shown that institutions with integrated and balanced performance management systems achieve better organizational results than their peers (Lingle and Schiemann 1996; Ahn 2001; Ittner et al. 2003; Lawson et al. 2005; Said et al. 2003; Sandt et al. 2001). Today, virtually all successful companies place a strong emphasis on, and have adopted, some type of performance management activities for the promise of quality improvement. A few examples of industries that have invested heavily in their performance management systems include the chemical industry, the financial industry, and the retail industry (Jovasevic-Stojanovic and Stojanovic 2009; IBM 2009; Mullie and Hill 2004).

Performance management, as it pertains to the AMP model, provides direct evidence that an organization can substantively understand and effectively reduce its operational risks to an acceptable level. A fully functioning biorisk management system (or any other management system) will be critically impaired if performance evaluation is absent or only partially implemented. The primary goal of performance evaluations is to ensure that the implemented mitigation measures are indeed reducing or eliminating risks. Performance evaluations also help to highlight biorisk strategies that are not working effectively. Measures that are not effective or are shown to be unnecessary can be eliminated or replaced. It may also be appropriate to reevaluate the overall mitigation strategy.

In contrast to mitigation, performance has historically received the least amount of attention in traditional biosafety systems. The reasons for this are varied. As in any management system, effective implementation of performance assessments requires a strong foundation of organizational sophistication and leadership, dedicated resources, and specific training. It also requires a quality method of determining success or failure. Measuring performance is not a short-term goal or something that can be purchased, but rather a long-term and evolving goal—it is an iterative process that must be continually evaluated and adjusted over time. Most importantly, management must be wholly committed to continual evaluation and improvement. There are often inherent barriers to overcome in the organization when implementing performance management tools and strategies. For example, common challenges include lack of an organizational learning culture, personnel resistance to change, distrust of management, and lack of motivation.

Yet, when implemented successfully, performance management can be one of the most powerful interventions an organization can embrace. Indeed, numerous fields and industries have used performance management to increase their productivity, reallocate resources more effectively, and increase profits. The healthcare industry has used performance management in various operational areas. For example, in 2012, the Virginia Department of Health developed a performance improvement plan that identified multiple strategies to save more than US$1.2 million annually on information costs, and created a more efficient online system that reduced redundant administrative processes and costs, and increased enrollment in a Medicaid planning program by 32% (US Centers for Disease Control and Prevention 2012).

LABORATORY BIORISK MANAGEMENT GUIDELINES

To better reduce biorisks in the laboratory and to standardize biorisk management approaches, the international community developed the CEN Workshop Agreement—Laboratory Biorisk Management (CWA 15793:2011). The CWA uses a management system approach—a framework that integrates best practices and procedures, and helps ensure that an organization can effectively achieve all of its objectives. CWA 15793 increases awareness of biosafety and biosecurity risks, and improves international laboratory collaboration and safety harmonization. It can also serve as the basis for new or revised legislation or regulations, and could, eventually, support laboratory certification/accreditation, and audits/inspections.

CWA 15793 is an objectives-based document that relies on acceptable best practices. It is not country specific, although it is designed to complement other management standards (such as ISO 9001, ISO 14001, and OHSAS 18001). It is important that organizations and other users recognize that CWA 15793 is not a technical specification. Rather, it is a performance-oriented document that describes what needs to be achieved; each organization must decide how to meet those expectations. Moreover, compliance to this document is completely voluntary. CWA 15793 is a seminal international biorisk management document, and although it does not define the AMP model explicitly, all of the components of the AMP model are addressed in CWA 15793. As such, CWA 15793 will be referred to in numerous chapters throughout this book.

PLAN-DO-CHECK-ACT

When management recognizes that operations are not running smoothly and laboratory errors are happening too frequently, it is useful to have a process in place to find a solution. In the 1950s, W. Edwards Deming proposed that the necessary processes needed to make a change or solve a problem within an organization should be managed in a simple continuous feedback loop. This process was illustrated and put into a cyclic and iterative four-step diagram, called the plan-do-check-act (PDCA) model, which is now commonly referred to as the Deming cycle or the Deming wheel (Deming 1950), as seen in Figure 2.2.

Biorisk management systems commonly depend upon the PDCA cycle; even CWA 15793 is built around the PDCA model. Most efforts to implement CWA 15793 in laboratories have focused on the PDCA model for just the performance aspect of the AMP model. However, PDCA can and should be used for the assessment and

FIGURE 2.2 The PCDA cycle.

mitigation components of the AMP model as well. Successful implementation of PDCA can lead to measurable improvements in an organization's efficiency, effectiveness, and accountability, among other indicators of quality.

The steps in the PDCA cycle are:

1. **Plan:** Plan a change and develop goals. In this step, one should think about where the organization is and where it needs to be. Objectives and processes should be established to meet targets and goals.
2. **Do:** Implement the plan, execute the process, and test the change.
3. **Check:** Study the actual results and compare them against the expected results. Review the test, analyze the results, and identify what has been learned. Measure performance. Assess how the risks are being controlled and if aims are being achieved.
4. **Act:** Request corrective actions to address differences between actual and planned results. Analyze differences to determine causes. Take action based on what has been learned. Use what has been learned to plan new improvements, beginning the cycle again. Review performance. Take action on lessons learned.

Chapter 8, "Evaluating Biorisk Management Performance," provides additional insight into applying the PDCA model. It is critical to work through each phase of the PDCA cycle for an effective performance improvement process. Similar to the assessment and mitigation phases, performance should not be driven or executed solely by a laboratory's biorisk management advisor or biosafety professional. As discussed earlier, quality performance in an organization relies on the culmination of input and support from numerous people in the laboratory or facility. Evaluating performance should not be only an administrative or management function. It is important to assemble and recruit a biorisk management team who will participate and help to develop and execute a performance improvement plan.

Further, all results of the performance process should be shared with the organization. The organization's areas of weakness, and its new goals and objectives, should always be communicated to staff to support the development of a collective, team approach. If the staff is unaware of organizational changes, or they are resistant to embracing the changes, the effectiveness and success of performance management will obviously be limited.

While the PDCA model is routinely used in management systems, it is most effective to use it as a discrete process within each step of the broader AMP model. Thus, it is vital to have a PDCA cycle within risk assessment, another cycle for implementation of specific mitigation measures, and a third cycle within the performance step.

CONCLUSION

Every organization that embraces biorisk management should employ the AMP model. The three components of the AMP model are (1) assessment, which identifies and prioritizes the biorisks present in the laboratory or organization; (2) mitigation, which applies various measures to reduce those risks; and (3) performance, which evaluates how well the organization reduces the identified risks. The AMP model is not linear; in other words, after completion of A, M, and P, an organization should repeat the process as often as necessary to maximize effectiveness. If there is shared commitment and determination to reduce biorisks and improve safety and security, the AMP biorisk management model will have profound effects on the organization's productivity, effectiveness, and safety. The following chapters on risk assessment, mitigation, and performance provide more detailed information about each component of the AMP model.

REFERENCES

Ahn, Heinz. 2001. Applying the Balanced Scorecard Concept: An Experience Report. *Long Range Planning*, 34: 441–461.

Deming, William Edwards. 1950. Elementary Principles of the Statistical Control of Quality. *JUSE*, 134–c6 (out of print).

Environmental Protection Agency. 2004. An Examination of EPA Risk Assessment Principles and Practices. http://www.epa.gov/osa/pdfs/ratf-final.pdf (accessed August 2014).

European Committee for Standardization. 2011. CEN Workshop Agreement (CWA) 15793— Laboratory Biorisk Management.

IBM. 2009. Banking Performance Management: Three Ways Banks Are Winning with Performance Management. http://www.ibm.com/us/en/ (accessed August 2014).

International Civil Aviation Organization. 2013. State of Global Aviation Industry. http://www.icao.int/safety/State%20of%20Global%20Aviation%20Safety/ICAO_SGAS_book_EN_SEPT2013_final_web.pdf (accessed August 2014).

Ittner, Christopher D., D.F. Larcker, and T. Randall. 2003. Performance Implications of Strategic Performance Measurement in Financial Services Firms. *Accounting, Organizations and Society*, 28: 715–741.

Jovasevic-Stojanovic, Milena, and B. Stojanovic. 2009. Performance Indicators for Monitoring Safety Management Systems in the Chemical Industry. *Chemical Industry and Chemical Engineering Quarterly*, 15: 5–8.

Kaplan, Stanley, and B.J. Garrick. 1981. On the Quantitative Definition of Risk. *Risk Analysis*, 1: 11–27.

Lawson, Raef, W. Stratton, and T. Hatch. 2005. Achieving Strategy with Scorecarding. *Journal of Corporate Accounting and Finance*, 16: 63–68.

Lingle, John H., and W.A. Schiemann. 1996. From the Balanced Scorecard to Strategic Gauges: Is Measurement Worth It? *Management Review*, 85: 56–61.

Loeb, Jerod M. 2004. The Current State of Performance Measurement in Health Care. *International Journal for Quality in Health Care*, 16: S1i5–S1i9.

Mullie, Christine, and R. Hill. 2004. Performance Management in the Retail Industry: Achieving Superior Corporate Performance. *Business Objects*. http://www.businessobjects.com/pdf/partners/ibm/ibm_owdretail_perf_mgmt.pdf (accessed August 2014).

Public Health Agency of Canada. Pathogen Safety Data Sheets and Risk Assessment. http://www.phac-aspc.gc.ca/lab-bio/res/psds-ftss/index-eng.php (accessed August 2014).

Said, Amal A., H.R. Hassabelnaby, and B. Wier. 2003. An Empirical Investigation of the Performance Consequences of Nonfinancial Measures. *Journal of Management Accounting Research*, 15: 193–223.

Sandt, J., U. Schaeffer, and J. Weber. 2001. Balanced Performance Measurement Systems and Manager Satisfaction. Otto Beisheim Graduate School of Management.

US Centers for Disease Control and Prevention. 2012. Public Health Practice Stories from the Field: National Public Health Improvement Initiative in Virginia. http://www.cdc.gov/stltpublichealth/phpracticestories/pdfs/PHPSFF_Virginia_v2.pdf (accessed August 2014).

US Department of Health and Human Services. 2009. *Biosafety in Microbiological and Biomedical Laboratories*. 5th ed. http://www.cdc.gov/biosafety/publications/bmbl5/bmbl.pdf (accessed August 2014).

World Health Organization. 2004. *Laboratory Biosafety Manual*. 3rd ed. http://www.who.int/csr/resources/publications/biosafety/en/Biosafety7.pdf (accessed August 14, 2014).

World Health Organization. 2010. Biorisk Management Advanced Trainer Programme. http://www.who.int/ihr/training/biorisk_management/en/ (accessed August 2014).

3 Risk Assessment

Susan Caskey and Edgar E. Sevilla-Reyes

ABSTRACT

The intent of this chapter is to define the goal of a risk assessment, and to explain in general terms how to conduct a risk assessment. This chapter also provides some concepts to consider when determining whether or not to accept some level of risk, and how to ensure that a risk assessment fits into a biorisk management system. The purpose of this chapter is not to serve as a detailed technical manual on biological risk assessment; rather, it provides the reader with the basic guidance required to understand and value a structured risk assessment approach.

DEFINITION OF RISK

A broad spectrum of risks may be present in any operation, including risks to the individuals directly involved in a particular situation, risks to the surrounding community, and risks to the environment. Successfully mitigating these risks requires first understanding them. Understanding the risk is the primary goal of a risk assessment. Risk is generally defined as the possibility that something bad or unpleasant (such as an injury or loss) will happen. More specifically, risk is the *likelihood that an adverse event involving a specific hazard or threat will occur and the consequences of that occurrence.* Simply, risk is a function of likelihood and consequences.

A hazard is something that has the *potential to cause harm.* In order for there to be a risk, there must be a situation for the hazard to cause harm. For example, a sharp needle is a hazard, but if the needle is in an empty laboratory, and no one is using it, there is no risk that someone will be injured by that needle.

Threats and risks are two terms often used interchangeably. A threat is someone with *intent to cause harm.* In order for there to be a risk, there must be a situation in which the threat can cause harm. For example, a criminal who aims to steal a computer is a threat; however, if there are no computers in the building, there is no risk of the criminal stealing one from that building.

Risk is always dependent on the situation.

At the most basic level, assessing a risk involves answering the following questions (Kaplan and Garrick 1981):

1. What can go wrong?
2. How likely is it and how likely are we to see it coming?
3. What are the consequences?

The goals of risk assessments are to allow for a better understanding of the risks, to determine acceptability of the risks, and to help define strategic risk mitigation measures. A structured and repeatable risk assessment process will allow these goals to be achieved.

In addition to the risk of exposure and infection for the personnel in the facility, and the potential for a release to the community and the environment, facilities that work with biological agents also may have risks associated with the loss or theft of intellectual property, physical property, or the biological agents themselves. All of these risks involving biological agents are collectively called biorisks.

BIOSAFETY RISKS

Biosafety risks are a type of biorisk that can affect humans, animals, or the environment after an accidental exposure or release of a biological agent. Personnel who work directly with biological materials face numerous biosafety risks; biosafety risks also exist for personnel who work indirectly with, or in close proximity to, infectious agents. Likewise, experimental animals exposed to infectious agents in a laboratory have the potential to expose and infect other animals housed in the facility. The public and agricultural community outside the laboratory or facility may also face biosafety risks in the event that an agent is unintentionally released into the environment.

While work with any infectious agent or toxin carries a biosafety risk, the severity of the overall risk is dependent upon a number of factors. These factors include properties of the biological agent (the hazard), properties of the potential host, and the work practices and procedures used when handling the agent in the laboratory.

BIOSECURITY RISKS

Biosecurity risks are a type of biorisk that results from a person who has malicious intent and has potential access to a hazardous material or facility. These risks primarily focus on theft of a biological agent, equipment, or information, but can also include misuse, diversion, sabotage, unauthorized access, or intentional release. The overall biosecurity risk varies with the intent and capabilities of the person or persons who want to conduct the malicious act.

In assessing a biosecurity risk, the malicious intent is typically focused upon an item of value, or asset, within the laboratory, and the threat's intended goal for that asset. In a biosecurity risk assessment, it is critical to define what assets exist within the facility. Once the assets are identified, a biosecurity risk can be defined as the likelihood that an asset would be stolen from the laboratory, and the consequences of loss (to include misuse of the asset following the theft). In contrast to many biosafety risks, biosecurity risks are often difficult to identify and characterize because they are dependent upon intent of the individuals and their level of determination to obtain or use the asset.

TRADITIONAL RISK ASSESSMENT APPROACHES FOR BIORISKS

Many experts who conduct biorisk assessments have traditionally evaluated agents based on predetermined risk groups. These agent risk groups consider an agent's ability to cause infectious disease, its transmissibility, and the availability of prophylactic and treatment measures (American Biological Safety Association 2014). Agent risk groups do not reflect the risk of accidental release or exposure in the laboratory. For example:

WHO Risk Group 2 (moderate individual risk, low community risk): A pathogen that can cause human or animal disease but is unlikely to be a serious hazard to laboratory workers, the community, livestock, or the environment. Laboratory exposures may cause serious infection, but effective treatment and preventative measures are available and the risk of spread of infection is limited. (World Health Organization 2004; American Biological Safety Association 2014)

Different national and international institutions have developed their own schemes for defining agent risk groups (European Agency for Safety and Health at Work 2014). Ideally, the risk assessor should use this information as only a starting point in the evaluation of the risk of a laboratory procedure. Unfortunately, many risk assessments are often based upon only the risk group classification, and there is minimal consideration regarding the laboratory processes or risk mitigation measures that are in place (US Centers for Disease Control and Prevention 2014).

For biosecurity risk assessments, many institutions rely upon the same risk group classifications to determine levels of security. Risk group classifications fall short of adequately defining biosafety risks; these classifications also do not provide the proper basis for conducting biosecurity assessments. This is especially true in light of the rapid expansion in the number of high-containment research facilities and the increasing amount of work conducted with dangerous biological agents. For this reason, many leading international biosafety experts have called for the development of a structured, quantitative biorisk assessment methodology (Wagener et al. 2008).

RISK GOVERNANCE AND BIORISK MANAGEMENT

Many different methodologies (or schemes) are used in the calculation of risk. For example, a probabilistic assessment establishes probabilities of occurrence, a relative assessment provides a comparison to other risks, and a dynamic process allows for decisions at each step to alter the final risk (Ezell et al. 2000, 2001, 2010). This chapter does not define which of these (or the many other) methodologies is best suited for conducting a biological risk assessment. Instead, it provides an overall structure by which the biorisk assessment should be conducted. The outline discussed here aligns with the principles of risk governance articulated by the risk and decision community (International Risk Governance Council 2005).

RISK ASSESSMENT METHODOLOGY

A risk assessment should be a structured process to identify and manage all of the risks that an entity faces. A structured process allows for better communication of the risks, comparability between risks, and accurate reassessments to identify any changes in the risk. A risk assessment reviews all aspects of the situation, including location, proposed work activities, personnel, storage, transfer and transport, destruction, access, and security, among others.

The principles of risk governance articulate that a risk assessment should be based upon the following three general steps:

1. Define the situation	What work is occurring?
2. Define the risks within the situation	What can go wrong?
3. Characterize the risks	How likely is it to happen?
	What are the consequences?

Following the characterization of the risks, management should determine whether or not each type of risk is acceptable. If a risk is not acceptable, then management must implement risk mitigation measures to lower that risk to an acceptable level. Risk assessments should be conducted periodically, and when any aspect of the work situation changes. Thus, the risk assessment process is continuous.

This chapter provides guidance on how to conduct a risk assessment for biorisks, specifically focusing on biosafety and biosecurity risks. Because the objectives of laboratory biosafety and biosecurity are different, the risks will be different, and, as a result, must be assessed independently. Although there are many similarities in the possible consequences for safety and security assessments, the initiating events—accidental or intentional—are different, and thus the likelihood of those risks need to be assessed differently (Snell 2002).

Biological safety and biological security risk assessments follow the general process as shown in Figure 3.1.

For a *biosafety risk* assessment at a facility, risk varies with:

- The properties of the biological agent (e.g., its physical state), the at-risk hosts (humans, animals, environment), and the specific laboratory processes, including any mitigation measures already in place
- The severity of the consequences to a lab worker or to the community and environment if there is a release

For a *biosecurity risk* assessment at a facility, risk varies with:

- The likelihood of successful theft from the facility of biological material, equipment, or information by an adversary (threat)
- The severity of the consequences of the theft based upon the properties of the asset stolen and the intent of the adversary

The risk assessment process starts by thoroughly defining the situation. This can be a formalized process that includes the identification of all the hazards and threats, work and target locations, proposed work activities, personnel and access levels, storage security, transfer and transport, destruction, and facility security, among others. This step can be accomplished at an informal level by simply describing who, what, and where in an outline format. Some examples of defined risks are presented in the biosafety and biosecurity risk assessment sections below. The assessment

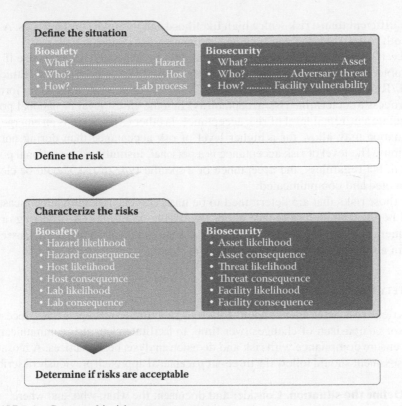

FIGURE 3.1 Common biorisk assessment process.

needs to identify and define all reasonable risks and clearly describe *what can go wrong* based upon the situation.

For each defined risk, the risk must be characterized to determine *how likely is the risk* to happen, and *what will be the consequences* if it does. To determine the likelihood of a risk, the risk assessment team must consider the factors that precede the incident. The incident is the point at which the risk occurs. To determine the consequences, the risk assessment team must consider the factors that occur following the incident (Figure 3.2).

Characterizing the risk requires a consideration of both the likelihood and the consequences. A risk in which there are notable consequences but a minimal likelihood

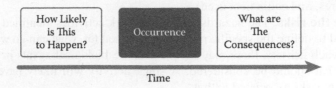

FIGURE 3.2 Relationship between likelihood, consequences, and time.

is very different than a risk with a high likelihood and minimal consequences. A risk with both high likelihood and high consequences is, of course, a high risk.

Once the risk has been characterized, it must be evaluated to determine if it is acceptable. More discussion on risk acceptance will be included in the "Ethics of Biorisk/Risk Acceptability" section of this chapter. In general, there is no formalized process to determine risk acceptability. In some cases, legal or national policy may dictate a minimal level of risk acceptance. In other cases, such as an emergency, the situation may allow for a higher level of risk acceptance than during normal operations. The level of risk acceptance is a personal, institutional, or national policy decision, but regardless, the acceptance of a specific type of risk should be clearly documented and communicated.

For those risks that are determined to be unacceptable, risk mitigation measures should be implemented to reduce either the likelihood of the risk occurring or the consequences of it occurring—or both. Biorisk mitigation measures are discussed in detail in a subsequent chapter of this book (Chapter 5).

BIOSAFETY RISK ASSESSMENT

A biosafety risk assessment should follow a structured and repeatable process to allow for comparison of changes over time, to facilitate clear risk communication, and to ensure compliance with risk and decision analysis best practices. A biosafety risk assessment should follow the three-step technical approach previously described.

1. **Define the situation.** Consider and document the what, who, and where.

   What—Identify the hazards. Hazard identification is a critical step of the biosafety risk assessment process. The risk assessment team must identify the biological agents to be handled (or, if the biological agent in the sample is unknown, the agents suspected to possibly exist).

   Who—Evaluate at-risk hosts. The risk assessment team must also consider the host range for the hazards identified. These hosts may include humans inside and outside of the laboratory, and extend to the agricultural and animal species outside of the laboratory in the event of an accidental release (the endemicity of the agent in the area should also be considered).

   Where—Define the work activities and laboratory environment. In defining the work activities, the risk assessment team must articulate and document the laboratory processes (including locations, procedures, and equipment used).

2. **Define the risks.** The hazards, hosts, and work activities identified should be used to define the specific risks to be assessed (or, what can go wrong?). From each activity, there may be one or more biological agents or procedures that should be considered. A single activity will likely have many different risks associated with it.

   Defining the risks must include a review of how individuals inside and outside the laboratory may be exposed to the hazard(s). These risks include the following examples:

- Risk to individuals in the laboratory (*laboratory workers*) of an infection:
    - Via droplets or droplet nuclei that have entered the upper or lower respiratory tract (inhalation route)
    - Through compromised skin or direct injection into the bloodstream (percutaneous route)
    - Through exposure to the mucosal membranes (mucosal route)
    - Via contact with the gastrointestinal tract (gastrointestinal route)
- Risk to an individual(s) outside the laboratory (*the human community*) of an infection:
    - Via droplets or droplet nuclei that have entered the upper or lower respiratory tract (inhalation route)
    - Through compromised skin or direct injection into the bloodstream (percutaneous route)
    - Through exposure to the mucosal membranes (mucosal route)
    - Via contact with the gastrointestinal tract (gastrointestinal route)
- Risk to animals outside the laboratory (*the animal community*) of an infection:
    - Via droplets or droplet nuclei that have entered the upper or lower respiratory tract (inhalation route)
    - Through compromised skin or direct injection into the bloodstream (percutaneous route)
    - Through exposure to the mucosal membranes (mucosal route)
    - Via contact with the gastrointestinal tract (gastrointestinal route)
- Risks to humans and animals as a result of a secondary exposure

After defining the various possible risks, the risk assessment team should determine if one general assessment will cover all the risks, or if multiple assessments are required. It is critical to recognize that the more detailed and specific the assessments are, the more useful the results will be in making risk mitigation decisions.

3. **Characterize the risks.** Risk, as defined previously, is a function of both likelihood and consequences. To determine the risk, the risk assessment team must answer: *How likely is it to happen? What are the consequences?* All the elements that influence the likelihood of infection and the likelihood of exposure should be combined to characterize the overall likelihood. Similarly, the risk assessment team must combine all the elements that define consequences of disease, following an exposure to a host, to determine what the consequences will be. These elements can be combined mathematically, semiquantitatively, or qualitatively (as high, medium, or low). Whichever process is used, it must be used consistently and it must be clearly documented.

    Hazard assessment. In a biosafety hazard assessment, the risk assessment team must consider those properties of the hazard that would influence its potential (or likelihood) to cause an infection. For a biological hazard, these would include such properties as the routes of infection of the biological agent in the laboratory and the natural environment, as well as the agent's infectious dose.* The routes of infection in the laboratory should include inhalation, ingestion, injection (into the bloodstream), and through mucosal membranes. The natural routes of infection include vector-borne, sexual transmission, and vertical transmission. These are important in assessing the risk to the human or animal community outside of the laboratory, and also in assessing the potential for a secondary transmission.

A hazard assessment would also review the properties of the disease(s) that would result from an infection caused by the biological agent (the consequences of the disease). This should include considerations of the morbidity of the disease within an individual, the morbidity rates (how rapidly the disease can spread), the mortality rates, treatment or prophylaxis options, species affected, and economic impacts for diseases that affect animals, etc.

Host assessment. It may not be necessary to assess the potential for an individual person or animal to be infected by the biological agent, as most biosafety decisions are defined in general terms and not designed uniquely for each individual. However, if individuals have a medical condition or are susceptible to disease because of a weakened immune system, or if the biological agent is unique to the region (e.g., the biological agent does not currently exist in the laboratory's external environment), these factors must be documented within the risk assessment process. Further, if a host assessment is deemed necessary, the risk assessment team must consider factors that might influence an individual's potential (or likelihood) of developing an infection, or would influence the potential (or likelihood) of the biological agent establishing a reservoir within the community (environment). It may also be necessary to consider the potential consequences of the disease to specific individuals or to host species within the environment.

In conducting a host assessment, the likelihood of infection and the consequences of disease can be characterized qualitatively (as high, moderate, or low), semiqualitatively, or mathematically. Again, it is important to document the process used and to be consistent.

How—Work activities and laboratory environment assessment. To assess the biosafety risks of the laboratory environment, the risk assessment team should review the types of laboratory processes performed, and identify any potential areas where an exposure to the hazard (biological agent) might occur. In addition, the risk assessment team must document and consider any existing biosafety measures that are in place to reduce this exposure. It is recommended that the team document the potential sources of an exposure separately from the existing

---

* It is especially important to consider agents with a low infectious dose, as they present a greater risk of infection; hazardous agents with a higher infectious dose typically do not warrant similar precautions.

mitigation measures. This allows for a better understanding of how the biosafety mitigation measures directly address the potential (or likelihood) of an exposure to occur.

The team should also consider the consequences to the laboratory in the event of an exposure that leads to a disease. In addition to the direct health impacts to potential hosts, consequences could include the loss of a program or institutional funding, or perhaps even a temporary suspension or termination of all research if the laboratory is found to be negligent. Such additional consequences should be considered and documented.

Overall risk characterization. To characterize the overall biosafety risks, the risk assessment team needs to compare the likelihood and the consequences of infection—either qualitatively or quantitatively, or using a hybrid process.

- Likelihood:
    - Likelihood of infection from the biological agents
    - Likelihood of an exposure based upon the work practices
- Consequences:
    - Consequences of infection/exposure to an at-risk host

Management Determination of the Risks

The risk assessment team, working with management and other stakeholders, should determine if the assessed risk is acceptable to the institution, individuals working in the institution, and the community. For some situations, the minimal level of acceptable risk may be defined by national or regional policy. The risk assessment process for determining that a risk is acceptable must be fully documented. For a risk that is determined to be unacceptable, the risk assessment team, management, and other stakeholders must identify the mitigation measures that are necessary to implement. Risk mitigation measures should be targeted to reduce the highest (or most unacceptable) risks. A subsequent risk assessment should be conducted after those measures have been implemented to ensure that the biosafety risks have been reduced to an acceptable level.

BIOSECURITY RISK ASSESSMENT

A biosecurity assessment includes defining the laboratory assets, threats, and facility vulnerabilities, as well as the current biosecurity program in place to mitigate biosecurity risks, and the impact or consequences of theft or destruction of the defined assets. Determining the potential security risks based upon these factors is the first step in implementing a biosecurity program.

Similar to the biosafety risk assessment, a biosecurity risk assessment should follow a structured and repeatable process that clearly defines the likelihood of targeting assets from the laboratory, the likelihood of an adversary (or threat) successfully acquiring the target, the institutional environment, and the consequence of a successful acquisition (and potentially subsequent misuse) or destruction of the asset.

1. **Define the situation.** Consider and document the what, who, and where.

   What—Identify and define the assets. The risk assessment team must first identify and then define and document the facility's assets that should be protected. This may include secondary targets, such as a backup generator used to provide cooling for critical storage units. A biological facility will likely have a variety of asset types. Assets include anything of value to the institution or an adversary. Examples of assets may include valuable biological material, such as pathogens and toxins, valuable equipment, intellectual property, or other sensitive information, reagents, and laboratory animals.

   Who—Define the threats. The team must then identify and evaluate the potential adversaries who may pursue those assets. A thorough threat assessment should include a consideration of adversarial types and capabilities, motives, means, and opportunities. It should consider adversary scenarios, as well as the likelihood of attack.

   Examples of adversarial types that could target assets at a biological facility include competitive researchers, criminals looking for items to sell, disgruntled employees, a terrorist organization, and animal rights activists.

   Where—Define the facility and laboratory security environment. In defining the environment, the risk assessment team should consider the vulnerabilities of the facility that contains the assets. It should also review the work performed in the laboratory, and who has access to the laboratory and its assets.

2. **Define the risks.** From the list of defined assets and threats, the risk assessment team can construct a series of potential risk scenarios based upon how and why an adversary may attempt to acquire (and possibly misuse) or attempt to destroy an asset. Risks could include the following examples (the specific risk defined should be unique to the biological institution):

   • Risk of an unauthorized person stealing valuable biological material for malicious use:
     – Example: A farmer intent on infecting a competitor's flock of birds.
   • Risk of an authorized person stealing valuable biological material for malicious use:
     – Example: An employee upset with a spouse and intent on making him or her sick.
   • Risk of an unauthorized person stealing valuable biological material for personal gain:
     – Example: A criminal intent on stealing and selling biological material or equipment.
   • Risk of an authorized person stealing or destroying valuable biological material for personal gain:
     – Example: An adversary intent on damaging a research project so that he or she may publish a similar research study first.

- Risk of an unauthorized person stealing equipment:
  - Example: A criminal intent on stealing a computer to subsequently sell it.
- Risk of an authorized person stealing equipment:
  - Example: An employee intent on stealing a refrigerator for personal use.
- Risk of an unauthorized person stealing an institution's intellectual property (in the form of information) or confidential information:
  - Example: A competitor intent on producing a competitive vaccine.
- Risk of an authorized person stealing an institution's intellectual property (in the form of information) or confidential information:
  - Example: A disgruntled employee intent on sabotaging an institution's reputation by leaking confidential information to the media.

Within each risk, there may be one or more assets that should be considered. The location of the assets (both long-term and while in transit) should be considered and documented. Based upon the variety of locations, and the implemented security and procedures, different assessments may be required for each location. The risk assessment team should determine if one general assessment will cover all locations or if multiple assessments will be required. Likewise, the risk assessment team must determine if generalized, notional threats can be used in assessing the risk, or if specific threats should be considered.

Some situations may not present a sufficient risk to warrant a full assessment. These may include low-risk assets, such as biological materials that are ubiquitous in nature or nonpathogenic to humans, or adversaries who are incapable or uninterested in theft. Eliminating unnecessary or unrealistic risks will help narrow the scope of the risk assessment to a more manageable size.

3. **Characterize the risks.** A biosecurity risk is defined as a function of the likelihood of targeting an asset from the laboratory or institution, the likelihood of successful theft (or acquisition) or destruction or damage of the asset, and the consequences of that theft or destruction/damage.

   Assets assessment. Based upon the defined risks, the risk assessment team should define *the likelihood of targeting the asset* by the relevant threat. Depending upon the asset, the defined likelihood will vary. For valuable biological assets, the uniqueness of the asset and any potential for misuse should be considered. For valuable equipment, the value and uniqueness of the equipment and any potential for misuse should be considered. For each of the defined assets, the team should review the various properties that make this asset attractive (or likely) to be stolen or destroyed by an adversary. Similarly, the team should consider what consequences to the facility, individual, or environment would result from the theft, theft and misuse, or destruction of this asset.

   Additionally, the team should also consider the consequences to the community of an accidental release indirectly caused by the theft or

destruction of an asset. This assessment should consider the potential
health impact to humans, the potential health impact to animal populations, the economic impact, the psychological and social impact, the
financial impact as a result of the loss of work or the cost of replacement, and the possibility of damage to the facility's reputation resulting
from a theft and misuse or destruction of the asset.

Adversary assessment. The risk assessment team should define the inten-
tions and access that the specific adversaries might have to each asset.
This is key for determining the vulnerability of the facility to the
adversary.

Adversarial types can be further categorized into persons with
authorized access to the laboratory or facility (insiders), and persons
with no authorized access (outsiders). There may be graded levels of
access to an asset; some insiders may have access to some, none, or all
of the assets.

Adversarial motivations to target these assets similarly vary, and
may include the intent to:

- Obtain financial gain
- Destroy proprietary information
- Cause a nuisance by damage or destruction
- Inflict casualties
- Spread fear
- Make a political statement
- Protest a specific activity
- Demonstrate frustration with management

Whenever possible, the risk assessment team should use the attributes of known adversaries when characterizing the threats to a facility.
However, this may be difficult if the existing threat environment is not
known or very little information is available. It may be necessary to
collaborate with security personnel or local law enforcement to obtain
this type of information. Alternatively, the risk assessment team can
create a set of notional adversaries whose attributes span the spectrum
of plausible adversaries for the facility. The notional adversaries could
be entirely theoretical, or could be based on existing data from the local
environment. The local law enforcement community is a good resource
to assist in this process, and may also be part of the security risk assessment team.

How—Facility vulnerability assessment. The risk assessment team
should assess *the likelihood of successful acquisition* of the asset based
upon the asset's location, the facility's vulnerabilities, and the capabilities of the adversary. For example, the risk assessment team should
consider this question: What are the possible facility vulnerabilities or
avenues that an adversary could exploit to gain access to the assets?
Answering this question will help determine the likelihood of success
for an adversary to steal or destroy specific assets.

Additionally, each of the five pillars of biosecurity should be evaluated in detail. The five pillars include:

1. Physical security (the physical protection of the building and its assets)
2. Personnel reliability (practices in place to grant access and authorizations of assets to employees)
3. Material control and accountability (inventory of assets)
4. Transportation (movement of biological materials and other valuable materials within and between institutions)
5. Information security (the protection of sensitive information)

Overall risk characterization. To characterize the overall risk, the risk assessment team should consider the likelihood that an adversary would target a specific asset and the likelihood of successful theft or destruction of that asset. In addition, the team should evaluate the consequences of that theft or destruction. These can be compared purely qualitatively or quantitatively, or using a semiquantitative process.

- Likelihood:
  - Likelihood of targeting the asset for theft or destruction based upon intent/motivation of adversary (or threat)
  - Likelihood of successful theft or destruction based upon facility vulnerabilities and the threat's capabilities
- Consequences:
  - Consequences of the theft or destruction of the asset

Management Determination of the Risks

The risk assessment team, working with management and other stakeholders, should determine if the assessed biosecurity risk is acceptable to the institution, individuals working in the institution, and the community. For some situations, the minimal level of acceptable risk may be defined by national or regional policy. For any risk that is determined to be acceptable, the risk assessment decision process should be well documented. For a risk that is determined to be unacceptable, the risk assessment team, management, and other stakeholders must determine which mitigation measures are appropriate to implement. Risk mitigation measures should be targeted to reduce the highest (or most unacceptable) risks. A subsequent risk assessment should be conducted after those measures have been implemented to ensure that the risks have been reduced to an acceptable level.

ETHICS OF BIORISK/RISK ACCEPTABILITY

How safe is safe enough? How secure is secure enough? Are these barriers robust enough to work with these specific risks? These are examples of questions that should be asked to determine whether the assessed risks are acceptable, tolerable, or unacceptable (intolerable). Often, risks have been perceived as unacceptable if the consequences are "catastrophic," even if there is only a small chance of that risk occurring; likewise, risks judged to have low consequences, even with a high likelihood, are almost always considered acceptable. Research shows that people use their personal knowledge and experience with a particular risk to make judgments about risk acceptability.

Contextual factors come into play when considering a group or an organization's culture. N.F. Pidgeon defines culture as "the collection of beliefs, norms, attitudes, roles, and practices shared within a given social group or population" (Pidgeon 1991). Differences in cultures lead to different levels of risk perception and risk acceptance. Recent events within the community, especially something covered by the local media, will also define the contextual factors regarding risk perception. Additionally, people seem to judge human-made risks very differently than natural risks, accepting natural ones more readily than self-constructed, human-made risks.

Formally, a categorization of risk should arise from decisions that involve risk ethics. A part of moral philosophy, risk ethics address the principles that morally guide rational choices on risk taking and risk exposure, which are key elements of the biorisk management system (Wanderer 2012). The following is a list of risk perception factors that should be considered when conducting a risk assessment:

- Individual factors:
  - Knowledge: Risk perception is often associated with available knowledge. Having a scarcity of knowledge on a subject may cause increased fear, while doing deep and extensive research on a subject may reduce the fear.
  - Demographic variables: For example, some communities may be more resistant to various industries than other groups.
  - Personality aspects: Risk-tolerant versus risk-averse personalities.
  - Familiarity and control: In general, people are willing to take further risks when they feel familiar with the situation, or when they feel that they are in control or gaining control of the situation.
  - Health (mental and physical).
  - Stress.
- Context factors:
  - Culture: Each social group has its own experiences, religion, beliefs, traditions, and mythologies, which will influence how each perceives a given risk.
  - Available alternatives.
  - Political context.
  - Economic factors, such as financial benefit for taking a risk, can change a person's perceptions, as well as the availability of resources.
  - Recent events.
- Characterization of the risk:
  - Likelihood.
  - Consequences.
  - Historical information.
  - Professional judgment.
  - Formal analysis (cost-benefit).
  - Positive aspects of risk.
  - Media information.

All of these factors may explain, for example, why two laboratories that work with the same biological agents and similar research programs may have divergent biorisk management strategies. For example, executive management may resist additional safety measures because of cost concerns, or they may demand additional safety measures to avoid legal sanctions. Employees who are familiar with the work may resist additional safety measures because they complicate the task at hand; alternatively, they may demand additional safety equipment to simplify their safety procedures. In all of these cases, a sound, defensible risk assessment may dispel unrealistic, skewed risk perceptions.

To ensure that a risk is acceptable to the relevant stakeholders, a structured communication process should be used to discuss and evaluate the risks, as well as to make risk mitigation decisions.

1. Relevant groups should be engaged early in the risk assessment process.
2. Possible options have to be identified; all options should be assessed, including their potential risks and benefits.
3. The worst possible risks should be eliminated or mitigated prior to proposing which risks are acceptable.
4. The relevant stakeholders should be allowed to discuss and consider the acceptance of the risks.
5. All stakeholders should agree on the risk acceptance decision.
6. Relevant stakeholders should be informed of implemented mitigation measures and changes in risks.

Risk communication serves to educate others about the risks and can help change attitudes. Good risk communication can help to alleviate fears regarding a risk or, inversely, can help to demonstrate that a risk is unacceptable. Risk communication should become part of a structured risk assessment process. More details on communication are presented in Chapter 9.

ROLES AND RESPONSIBILITIES FOR RISK ASSESSMENT

It is important to emphasize that the risk assessment process should not be driven or executed solely by a laboratory's biorisk management advisor. Rather, a quality risk assessment is the culmination of contributions from numerous people in the laboratory or facility.

- **Biorisk management advisors/biosafety professionals:** These individuals are members of the laboratory staff who provide advice and guidance for laboratory biorisk management issues and workplace risk assessments. These individuals facilitate the risk assessment by gathering pertinent information to define risk, and using that information to characterize risks in terms of likelihood and consequences. The biorisk professionals should act as communicators to link the hands-on laboratory staff and contractors with management and other stakeholders. They should be knowledgeable of laboratory activities, sources of potential exposure, and means of effective control. They should also act as consultants for recommending and implementing appropriate mitigation measures that result from the risk assessment. Further, the biorisk professionals should have the most extensive understanding of the results from the risk assessment.

- **Principal investigators/scientists/researchers:** These individuals provide the primary information and data for the risk assessment. They are also expected to ensure that risk assessments have been completed. They must understand the risk assessment results, and provide input to management regarding practical implementation of recommended mitigation measures. They are also responsible for ensuring that at-risk employees have been informed of the risk assessment results, and training them on the necessary risk mitigation measures. The understanding and support of a risk assessment by the scientific staff is critical for effective biorisk management.

- **Security and response personnel:** These individuals may also provide valuable insight into risk assessments. For example, outside agencies, such as local law enforcement, may have knowledge about potential threats present in the local community. Security force personnel may be involved in the implementation of biosecurity mitigation measures, or they may act as inspectors to check the functionality of the mitigation measures. Other specialty agencies may also be necessary for the biosafety risk assessments, such as a hazardous materials team, an incident response team, the local fire department, or other first responders.

- **Legal consultant or legal department/public relations/labor safety officer:** These individuals may or may not have any direct involvement with the technical risk assessment process. However, their role is instrumental in risk communication. Expert opinion from this group is valuable when mitigation measures and policy changes need to be circulated among laboratory workers and the general public in order to gain their understanding and support. Their opinion may also need to be considered during the risk prioritization process. As these individuals often are not familiar with a laboratory or laboratory biorisk management, the primary individual responsible for conducting the risk assessment, such as the biorisk management advisor, should also be involved to optimize communication and understanding.

- **Laboratory contractors, waste handlers, maintenance staff, and janitorial crews:** These individuals are directly affected by laboratory risks, and they often have limited knowledge about the hazards to which they are exposed. These individuals should be engaged regarding their concerns and their level of understanding of the risks involved, and how the results of the risk assessment will impact them. It is vital to gain their support for implementation of any mitigation measures.

- **Executive management:** These individuals, which may include laboratory directors and high levels of management, will generally not conduct or be directly engaged in the risk assessment process. However, because they are ultimately responsible for the organization's biorisk management system, it is absolutely critical that this group support (and if necessary, direct) laboratories to conduct risk assessments, including the allocation of staff time and resources to perform the necessary data collection and analysis.

Executive management will be ultimately responsible for building the infrastructure and capacity that, in turn, supports establishing precautions and standard operating procedures (SOPs) to minimize laboratory risks. Mutual understanding between the risk assessment team and executive management is essential for optimal risk assessment outcomes. Resource allocation and financial support from this group is necessary to conduct the risk assessment and implement the appropriate and necessary biosafety and biosecurity measures. Risk assessment results are often confusing; any encountered problems may be overcome by engaging in dialogue with executive management in the early stages of risk assessment. It is also essential that risk assessment results be written in clear and concise language to facilitate understanding by executive management.

- **Administration (business and logistical support persons):** These individuals have limited access to a laboratory area, but typically have daily access to the people who work there. These individuals generally have limited scientific knowledge; thus, technical or scientific assessment results and the subsequent policies must be communicated in a way to ensure that the salient points are understood. However, administration should be consulted and engaged more thoroughly in any assessment of risks directly related to information management under the purview of administrative professionals. Biosecurity measures may significantly affect this group (such as compiling and distributing documentation), and therefore will require their full support in making facility operational changes.
- **Community stakeholders:** Individuals within the community may or may not be engaged, depending on the level of interest in the laboratory's operations. However, it may be prudent and a good practice to inform all outside visitors and the family of laboratory personnel about any potential risks they may encounter and the protocols in place to keep them safe— specifically, how the risks have been effectively managed or controlled. It will also be important to communicate the general results of the risk assessment to community stakeholders to ensure that they are aware that the facility understands and controls its risks in a responsible way.

Once all of the risks have been identified and communicated, the relevant laboratory and institutional staff should work together toward efficiently controlling or reducing the risks to an acceptable level.

Stakeholders can work together to determine the necessary *biosafety* practices required to perform a task with a specific organism. Likewise, stakeholders can work to determine the necessary *biosecurity* measures required to ensure a proper level of security has been established. This should be done in close collaboration with other responsible institutional parties, such as an institutional biosafety committee (IBC) or biorisk management committee, the environmental health and safety department, the security department, any animal care and use committees, or any engineering or facility departments.

CONCLUSION

This chapter defined the goals of a risk assessment as a better understanding of the risks present at a facility, a determination of the acceptability or unacceptability of identified risks, and assistance to facilities working with biological agents in defining strategic risk mitigation measures. Using a structured and repeatable risk assessment process for both biosafety risks and biosecurity risks will support achieving these goals.

At the most basic level, assessing a biological risk involves answering the following questions:

- What can go wrong?
- How likely is it and how likely are we to anticipate it?
- What are the consequences?

Analyzing the risk is a function of the likelihood of the risk occurring and the consequences if it happens. Risk acceptability or unacceptability should be based upon considering such questions as:

- How safe is safe enough?
- How secure is secure enough?
- Are these barriers (security or containment strategies) enough to work safely with these risks?

The answer to all of these questions should be a joint effort among all the stakeholders in the laboratory and should involve interested parties within the community. In other words, the risk assessment process should be conducted by a team and evaluated by a team—and explicitly supported by executive management.

REFERENCES

American Biological Safety Association. 2014. Risk Group Classification of Infectious Agents. http://www.absa.org/riskgroups/ (accessed August 2014).
European Agency for Safety and Health at Work. 2014. Risk Assessment for Biological Agents. EU-OSHA. https://osha.europa.eu/en/publications/e-facts/efact53 (accessed September 2014).
Ezell, Barry C., S.P. Bennett, D. Von Winterfeldt, J. Sokolowski, and A.J. Collins. 2010. Probabilistic Risk Analysis and Terrorism Risk. *Risk Analysis*, 30: 2010.
Ezell, Barry, J. Farr, and I. Wiese. 2000. Infrastructure Risk Analysis Model. *Journal of Infrastructure Systems*, 6(3), 114–117.
Ezell, Barry C., Y.Y. Haimes, and J.H. Lambert. 2001. Risks of Cyber Attack to Water Utility Supervisory Control and Data Acquisition Systems. *Military Operations Research Journal*, 6(2), 30–46.

International Risk Governance Council. 2005. *IRGC White Paper No. 1: Risk Governance— Toward an Integrative Approach.* Geneva. http://www.irgc.org/IMG/pdf/IRGC_WP_ No_1_Risk_Governance__reprinted_version_.pdf

Kaplan, Stanley, and B.J. Garrick. 1981. On the Quantitative Definition of Risk. *Risk Analysis*, 1: 11–27.

Pidgeon, Nick F. 1991. Safety Culture and Risk Management in Organization. *Journal of Cross-Cultural Psychology.*

Snell, Mark K. 2002. *Probabilistic Security Assessments: How They Differ from Safety Assessments.* SAND Report SAND2002-0402C. Sandia National Laboratories.

US Centers for Disease Control and Prevention. 2014. Biological Risk Assessment Worksheet. http://www.cdc.gov/biosafety/publications/BiologicalRiskAssessmentWorksheet.pdf (accessed September 30, 2014).

Wagener, Stefan, A. Bennett, M. Ellis, M. Heisz, K. Holmes, J. Kanabrocki, J. Kozlovac, P. Olinger, N. Previsani, R. Salerno, and T. Taylor. 2008. Biological Risk Assessment in the Laboratory: Report of the Second Biorisk Management Workshop. *Applied Biosafety*, 13: 3.

Wanderer, Emily. 2012. Perspectivas antropologicas sobre el riesgo, la historia y la bioseguridad. Los Comites Institucionales de Bioseguridad y las Competencias Profesionales AMEXBIO. May.

World Health Organization. 2004. *Laboratory Biosafety Manual.* 3rd ed. Geneva. http:// www.who.int/csr/resources/publications/biosafety/WHO_CDS_CSR_LYO_2004_11/ en/index.html.

4 Facility Design and Controls

William D. Arndt, Mark E. Fitzgerald,
and Ross Ferries

ABSTRACT

The risk of working with hazardous biological materials requires the implementation of an effective biorisk management program dedicated to protecting laboratory personnel and preventing the accidental or intentional release or removal of hazardous materials from the laboratory. Understanding these risks prior to constructing or renovating a laboratory is vital to ensure the success of an effective biorisk management program once work has commenced. The design of the laboratory and the containment measures implemented can negatively impact a biorisk management program. Therefore, design decisions and selection of specific containment measures should be based on a comprehensive biorisk assessment as opposed to defaulting to predefined solutions that may not be suitable in all cases. This chapter will discuss how the design of laboratories and the chosen containment measures impact biorisk management. Particularly, this chapter will also introduce the concept of using a risk-based design strategy to assist with the selection of suitable biorisk mitigation measures, reducing the chance that the facility will be over-engineered or that valuable resources will be wasted.

INTRODUCTION

Accidental exposure to hazardous biological agents or toxins and their inadvertent release into the environment are inherent risks of working with harmful biological materials. To minimize these risks, persons within the lab and the surrounding community should rely on the implementation of an effective biorisk management program that uses both primary and secondary containment barriers to decrease the likelihood of such events occurring. While the main functions of primary and secondary containment barriers are discussed in Chapter 5, this chapter discusses in further detail the importance of secondary containment barriers, and ultimately how the design of the laboratory and the selection of mitigation measures can impact the establishment and sustainability of an effective biorisk management program.

Laboratories designed to handle and store hazardous biological materials are historically classified into one of four biosafety levels (BSL1 to BSL4) recognized by the US Centers for Disease Control and Prevention (CDC), the US National Institutes of Health (NIH), and the World Health Organization (WHO). These predefined solutions of ascending levels of containment, or biosafety levels, are based on a combination of facility design features, operational practices and procedures, and safety equipment employed to mitigate the biorisks associated with handling and storing a particular agent to an acceptable level. *Biosafety in Microbiological and Biomedical Laboratories* (BMBL) (US Department of Health and Human Services 2009) and the *Laboratory Biosafety Manual* (World Health Organization 2004) are two guidance documents that describe the main biosafety risk mitigation measures associated with each of the four biosafety levels. However, the guidance in these two documents has been widely interpreted, and implementation of these biosafety levels often varies significantly from facility to facility.

A process has not been standardized within the international community to identify under what conditions or environment it is acceptable to conduct work with hazardous biological materials. This is especially the case in settings where laboratories may have been designed with biorisk mitigation measures selected based solely on those defined for a specific biosafety level. Determining the biorisk mitigation measures required for conducting work with hazardous biological materials in as safe and secure a manner as possible relies heavily on performing an assessment that identifies potential risks associated with handling and storing a specific agent. Unless hazards are identified effectively, it is not possible to assess the risk associated with the facility and associated activities (European Committee for Standardization 2011). For this reason, it is imperative that ample support is given to the individuals responsible for performing the risk assessments, ensuring an accurate estimation of the potential risks is made. For additional information related to the biorisk assessment process, refer to Chapter 3, which describes the main goals and strategies for conducting biosafety and biosecurity risk assessments.

Traditionally, the strategy of performing a biosafety assessment and selecting the appropriate biosafety level, based on the identified risks, has been implemented effectively in many facilities; however, in some instances, utilizing this strategy can be a challenge. Very few facilities worldwide have all the necessary resources to operate under ideal conditions that would allow them to effectively implement and maintain the risk mitigation measures recommended in the BMBL and the *Laboratory Biosafety Manual* for each of the respective biosafety levels. A number of factors must be considered prior to selecting mitigation measures, such as deficiencies with the existing infrastructure, an absence of appropriate and reliable utilities, shortage of funds to purchase and maintain equipment, insufficient training on how to properly use and maintain equipment, or a general lack of knowledge on international biosafety and biosecurity best practices. For these reasons, it is important to understand that before deciding on the biorisk mitigation measures to address a facility's risks, management must have confidence that those mitigation measures are appropriate, necessary, and sustainable. Otherwise, resources may be misapplied, creating a facility that is neither sustainable nor useful to the end users. For instance, it might be suggested that a laboratory performing diagnostic testing for *Mycobacterium tuberculosis* should be constructed as a BSL3 lab, equipped with HEPA filtration on the exhaust air system to protect against the potential release of aerosolized agent. However, if the lab in question does not have sufficient funds to operate its air handling systems year-round, and does not have the funds required for periodic testing and replacement of the HEPA filters, then this approach would not be sustainable, the resources applied to implement this approach would ultimately be wasted, and it would fail to mitigate the risk as intended.

An alternative strategy would be to evaluate what specific mitigation measures to implement from another perspective. Instead of using a predefined solution, such as biosafety levels and prescriptive facility elements, facility owners and design teams should focus on the biorisk mitigation measures themselves. In any laboratory design, predetermined, tried, and tested methods can be applied with reasonable expectations for performance outcomes, but new strategies may be needed to address unique design drivers and sustainability challenges that are not best solved by traditional thinking. The design strategy should place a strong emphasis on identifying the specific mitigation measures required to reduce the risks associated with handling and storing infectious biological materials instead of implementing mitigation measures solely from a biosafety level or predefined solution perspective. This alternative risk mitigation-based strategy will decrease the likelihood of creating a situation in which a facility includes overcomplicated engineering elements or unsustainable systems.

DESIGN FOR BIORISK MANAGEMENT

Designing a laboratory to handle and store hazardous biological materials is a time-consuming and complex process involving numerous stakeholders with differing opinions on facility design outcomes and the equipment required to support the end users' needs. All too often, laboratories around the world are designed and constructed with a limited understanding of the end users, and how their needs will affect building usage upon completion of construction and in the future as science, mission, and technology change. Consequently, facilities may be overdesigned or inadequately designed, and not capable of supporting safe and secure operations over time because of flaws in the building layout or engineering systems, or there are insufficient resources to ensure equipment continues to perform as expected.

Overdesigned facilities can be attributed to an uninformed design caused by a failure to gather the necessary information that has the potential to impact the overall outcomes of the design. For example, the appropriate scientific, safety, or engineering personnel who will eventually occupy and maintain the facility are not consulted to effectively determine their needs and abilities. Rather extreme examples, encountered by the authors, are laboratories that have directional airflow problems that result in air flowing from potentially contaminated areas toward clean areas. In many cases, these projects have been executed by design teams or builders experienced with clean laboratories who assume that a laboratory for work with biological agents has similar requirements. The failure to research containment requirements, to study the design of similar facilities, and to consult appropriately informed safety personnel can lead to the construction of facilities with airflow problems, which are often very expensive to remedy and place workers at increased risk.

A lack of local knowledge and capabilities related to laboratory design best practices can also lead to improperly designed labs. This situation can occur when a design at a distance strategy is used to compensate for a lack of local expertise.

The most likely scenario for relying on this type of strategy is when facility owners believe or recognize that the expertise of local architects and engineers is not well suited to designing and constructing laboratories that will handle and store hazardous biological materials. In such cases, the facility owners often hire experts from distant locations to design and, in some cases, construct their laboratory. This can be an effective process if adequate time and funds are made available to ensure designers gather the level of detailed information required to thoroughly understand the end users' needs and how the building will be utilized once construction is completed and in the future as technology advances or processes change. However, when the design team and the users are not geographically co-located, it can be difficult to organize sufficient discussions to develop a design where protocols and facility features function smoothly with each other. For example, the protocols for entering and exiting large-animal holding rooms can be very different from one facility to the next. A designer in one country may assume users will remove disposable coveralls, place them in a bin for sterilization, and then shower out of the holding area. This would require a relatively small change area with a shower. However, the users of the facility may actually be accustomed to a protocol of wearing reusable rubberized boots, pants, coats, and hats, and will need a space to disinfect and store these items prior to showering. The space they need is much different than the designer would have assumed, instead requiring separate areas for removing and storing personal protective equipment (PPE), in addition to change rooms and showering facilities. Without the opportunity for detailed discussions, there may be deficiencies in the facility design that will ultimately compromise the quality, efficiency, safety, and security of the work.

Lastly, challenges have arisen when a prefabricated design is used. These projects typically suffer from similar problems as design at a distance, in that there may be a lack of communication between the designers and the users, as well as a lack of understanding of local environmental conditions by the designers. These projects typically prefabricate a finished laboratory and deliver the laboratory to the site in near-complete or semicomplete form. Prefabricated components manufactured in one location and transported to another climate or construction environment are often difficult to maintain and sustain by local laboratory personnel. Heat, humidity, dust, pests, and altitude changes can have dramatic effects on materials and operation of mechanical systems. Additionally, construction materials used for a prefabricated lab may be very different than those that local trades are accustomed to, making minor maintenance problems, such as fixing a door, patching and painting damaged finishes, or adjusting mechanical, electrical, and plumbing systems, quite significant. For instance, damage to steel casework may go unrepaired and become a safety hazard in the lab for a significant length of time if no local trades have the materials and skills to repair it, whereas locally fabricated wood casework might be easily repaired in one day.

A unique set of skills is required to design laboratories, and these skills are generally acquired only through experience. Scientists, medical professionals, and biological safety experts who have a good understanding of the biological agents they work with and biorisk management best practices do not typically gain experience in laboratory design principles and best practices through their normal educational process or daily activities. Similarly, most architects and engineers are not educated about the risks associated with different biological agents or toxins, and how these risks are mitigated or amplified by routine procedures. The work of designing a laboratory should be entrusted to a group of experts, with wide-ranging experience, rather than a few designers. The best facility designs generally come from a well-rounded team of architects and engineers, experienced in laboratory design, working closely with scientists, biological safety and security professionals, and maintenance personnel who have an understanding of how scientific processes influence, or are influenced by, the building systems and layout. It is often quite challenging to gather the right mix of personnel and create the ideal conditions for the design process, with time for multiple meetings with key stakeholders, peer reviews by experts in the field, and mock-ups of critical areas. Even when achievable, the process of developing a laboratory facility is expensive and time-consuming. Therefore, it is not surprising that labs are not always well designed, are fraught with operational problems, and often do not adequately address the biorisks associated with the infectious material routinely used in the facilities.

Effectively managing the biorisks present at a facility is highly dependent on understanding what primary and secondary containment measures are truly necessary, based on the risks identified, and how those containment measures work together with other biorisk mitigation strategies. Examples include the implementation of administrative controls, practices and procedures, and PPE to mitigate risks to an acceptable level. Primary containment barriers typically focus on biosafety cabinets (BSCs), enclosed containers, and other equipment designed to protect laboratory personnel from possible exposure to infectious materials (US Department of Health and Human Services 2009). Secondary containment barriers concentrate more on the design and construction of facility elements that contribute to protecting not only the laboratory personnel, but also the surrounding community and the environment outside the laboratory, from a potential accidental release of infectious material (US Department of Health and Human Services 2009). Depending on the biorisks present in the facility, these secondary containment measures can vary from simple measures, such as autoclaves and hand washing equipment, to more advanced measures, such as specialized ventilation equipment to maintain directional airflow in the facility and filtration of laboratory air prior to discharge into the outside environment. Conducting comprehensive biorisk assessments early, and throughout the design of a project, should guide all decisions on what types of containment measures to include in the facility.

RISK-BASED DESIGN DECISIONS

The initial phase in making risk-based design decisions relies on the completion of biosafety and biosecurity risk assessments, which identify, evaluate, and prioritize the risks present at the facility, and enable the key stakeholders to make the final decisions on the appropriate risk mitigation measures to be implemented at the facility. Chapter 3 emphasized the importance of identifying the hazards (biosafety) and threats (biosecurity), and understanding the likelihood and consequences associated with the risks prior to determining risk acceptability and deciding on the required mitigation measures. It is essential that every member of the design team and key stakeholders agree on the risks that need to be addressed in the overall facility design, and how best to address those risks.

In laboratories that work with biological agents and toxins, biorisks may stem from the following:

- **The nature of the agents or toxins present in the facility.** For instance, a low infectious dose requirement and the severe consequences associated with infection by most viral hemorrhagic fevers (e.g., Ebola virus) are such that any work with these agents will inherently pose a significant biorisk to laboratory personnel.

- **The scientific procedures carried out.** It is important to understand what procedures are routinely conducted since this can significantly affect the relative biosafety and biosecurity risks identified for an agent in a comprehensive biorisk assessment. For example, work conducted with an agent may be determined to represent a moderate risk, but the risk may increase in the event that larger amounts of the same agent are routinely handled, such as in vaccine production facilities, or if work activities change that could increase potential exposure, such as if aerosolization studies are to be conducted. Conversely, an agent that may present a relatively high risk under certain conditions may actually represent a much lower risk if smaller amounts or attenuated strains are primarily used, such as in a clinical setting.

- **The risk of exposure to staff working in the facility.** Some activities can increase the potential to aerosolize hazardous agents (e.g., robotic assays that handle many samples rapidly), and procedures or policies can unnecessarily place workers in contact with others who may be contagious.

- **The risk to the environment or persons outside the facility.** Handling hazardous biological materials can pose a great risk to the users as well as the surrounding community. It is the inherent characteristics of the agents and the procedures being conducted in the facility that will determine the likelihood of infection and extent of a release into the surrounding human and animal populations. This can be a greater concern when facilities routinely handle agents that are relatively stable in the environment, have a low infectious dose, and have the ability to be easily transmitted between hosts, as is the case for agents such as foot-and-mouth disease (FMD) virus, African swine fever virus (ASFV), *Mycobacterium tuberculosis*, influenza virus, and polio virus.

- **The risk of theft of biological agents or toxins.** An individual or group of individuals, either associated or not associated with the facility, may wish to steal materials to use themselves or to sell to others for the purposes of causing harm.

- **The risk to the facility and the users from outside threats.** An individual or group of individuals not associated with the facility may wish to sabotage the facility. Historical examples include animal rights activism or other types of government or ideological protests.

Once the risks associated with the work conducted at a facility are understood and characterized, it is necessary to evaluate and prioritize the risks to ensure safe and secure operations, as well as the best use of available resources. Determining whether a biosafety or biosecurity risk is high, moderate, or low, and whether the risk is acceptable or not, is a subjective process that can vary between individuals and facilities depending on local laws, culture, experiences, management perspectives, and even current events. However, understanding site-specific factors and available resources is an important consideration in risk assessments that are routinely overlooked when predefined solutions are used for selecting biorisk mitigation measures. For example, laboratories in remote locations, with a limited number of people who live in close proximity to the lab, may possess a good level of protection against outside security threats, and a lower relative risk of exposure to the surrounding community in the event of an accidental release from the facility. Conversely, more densely populated areas would theoretically pose a greater security risk to the facility, and increase the risk of spread into the surrounding community in the event of an accidental or intentional release. At the same time, a specific agent may also pose a very different risk depending on the location of the facility. This is especially the case when comparing the risks associated with an agent considered to be endemic in one area and nonendemic in another. An accidental release of the FMD virus in a region where the agent is common in nature, such as Sub-Saharan Africa, may not be as great a concern as it is in North America, where the virus is not normally present. Even if it is possible to address every risk inherent in the design project, it is important for the design team to understand which risks should be given top priority, based on their likelihood and consequences, in order to make well-informed design decisions and the best use of available resources.

With the risks well understood and prioritized, it is then possible to begin discussing mitigation strategies. In the design of a facility for work with biological agents, some of the mitigation strategies will be inherently based on administrative controls (policies, standard operation procedures, etc.), engineering controls (BSCs, waste treatment, etc.), and facility design features (layout, workflow, etc.). However, effective biorisk mitigation strategies must involve a combination of administrative and engineering controls and facility design solutions to mitigate the biorisks present and to ensure sustainable, safe, and secure operations.

THE DESIGN PROCESS

Funding streams and expenditure cycles vary tremendously depending on the country, institution, donor, or corporation that provides the financial support. For this and several other reasons, financial support for the design and construction of a new laboratory facility can be difficult to predict. In some cases, unallocated funds at the end of the year may suddenly become available for projects on the condition that the money is spent quickly or applied to specific activities. In other cases, funds become available but are allocated to specific individuals or companies to advance the design and construction of a project, yet they may lack or choose to exclude appropriate multidisciplinary expertise. Regardless, these funding sources and expenditure cycles can pressure design and construction schedules in ways that may adversely impact the potential for project success. Additionally, ministries, shareholders, donors, and other potential funders may not have the background to appreciate the necessary complexity of such facilities, and often expect immediate return on their investment. Pressure to begin construction immediately is understandable because design and documentation can be very abstract when compared to the tactile, tangible building elements that appear during construction. Hence, those who have limited experience with traditional design and risk assessment processes, particularly when applied to a complex biological facility, may wish to see physical evidence of their investment too early in the process. This is often extremely damaging to long-term project success. When concrete slabs are poured, stair towers are constructed, and lifts and other physical elements are in place, it becomes difficult (or even impossible), time-consuming, and expensive to adjust aspects of the design. However, due diligence in information gathering, stakeholder buy-in for critical decision making, and appropriate design evolution prior to the onset of construction can yield significant savings in both time and cost when appropriately accommodated in the planning process. This is true for project capital costs as well as long-term ongoing costs to operate and maintain the facility.

Figure 4.1 illustrates the typical phases of a laboratory design project (from pre-design and programming to occupancy), as well as how the impact of decision making changes as the project progresses to completion. Decisions to include or exclude certain spaces or features can be easily made early in the design process with little negative impact on the project schedule. At this point, good information and good ideas often positively impact the project. As the project progresses, it becomes more difficult and more costly to implement changes, and new ideas and information become harder to incorporate in a productive manner. By way of example, in the programming phase of a project, it is easy to decide to include a dedicated space for security personnel to inspect incoming packages. However, the decision to add such a space, once the building has been completely designed, or worse is already under

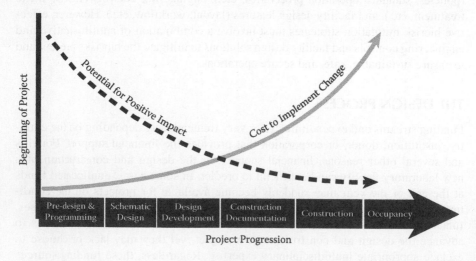

FIGURE 4.1 Decision-making cost and impact curve for project implementation.

construction, is generally a very expensive proposition. Multiple drawings will have to be revised by architects and engineers, and supporting equipment may need to be resized in order to support the additional air, plumbing, and power requirements. In addition, there will be costs for delaying the project, which will affect the contractor as well as the design team. As a result, it could be cost prohibitive to add this space to the design at that stage in the process.

PROJECT STAKEHOLDERS

A critical part of any successful project is including a balanced stakeholder group with a variety of opinions, perspectives, and skill sets. In order to effectively clarify roles and responsibilities, as well as facilitate decision making, the stakeholder groups should be divided into multiple tiers. A small core team of directors, key scientific staff, risk assessors, and funding agency representatives should be identified early in the process; this team should hold the final decision-making authority when competing opinions and unresolved issues present themselves. The core team should meet regularly to set and maintain progress toward project goals, monitor the design evolution, and ensure that outcomes are consistent with the overarching aspirations of the institution. The core team, however, should be a small subset of the complete project stakeholders who must be included in the design process. Representatives from individual scientific departments should be included in discussions with the design team to ensure that their essential needs are met. This group of users should include scientific staff who represent all programs, including laboratory directors and supervisors as well as technicians and other personnel who will use the lab on a daily basis. Other building users who should be included in the design process include operations and maintenance personnel, safety and security personnel, waste handlers, risk communication personnel, and administrative personnel. Assembling a comprehensive team of building users is essential, but including and communicating with individuals who will not directly work in the facility, but who can provide important perspectives, can also help ensure the overall success of the building project.

In addition to the aforementioned stakeholders who are directly connected to the project (i.e., work in or operate the facility or provide funding for the facility), there are often other groups who are peripherally connected to the project who will have concerns as well. Members of the surrounding community, for instance, may have concerns about the work going on in a laboratory, particularly if dangerous agents are present, and may need to be informed about the planned safety features of the lab in order to alleviate any concerns about their health or property being at risk. There may be regulatory or accreditation groups that will evaluate the facility against certain criteria or standards before granting approval for that facility to work with biological agents or animal subjects. Additionally, introducing the emergency response agencies responsible for responding during emergency situations (e.g., fire, criminal activity) to the facility and the activities to be conducted is vital to ensure the safety of all personnel who may be required to enter the facility.

While it is imperative to identify a core group that will have ultimate responsibility for making decisions, it is best when a laboratory design project can obtain

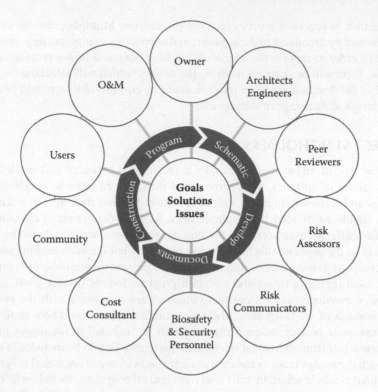

FIGURE 4.2 Integrated team of stakeholders, designers, and consultants.

input from the full range of stakeholders, as well as design experts and consultants, throughout the programming and design process. Though it may seem cumbersome to involve a large number of individuals in all discussions, including a wide range of perspectives is often the best insurance against errors or omissions that can result in difficult-to-operate facilities and costly, last-minute, or postconstruction changes. Figure 4.2 illustrates the concept of bringing together a wide range of stakeholders, design experts, and consultants who are focused on common project goals, issues, and solutions throughout all stages of the design process. Often it is advisable to involve peer reviewers as well, particularly when an institution has decided to change its working methods or the range of agents under study. Peers from similar institutions with similar work can be an invaluable resource in all stages of the design process.

PREDESIGN

Prior to the onset of the design phase, several project-related decisions can be made. These decisions establish boundary constraints for future work, and govern many of the early efforts that ultimately affect the final built environment. Ideally, owners will know the intended facility occupant(s), the mission of the laboratory units, the approximate size of the identified space needs, an order of magnitude estimate of the available funds for design, construction, and continued operations, the available site(s) for consideration during the predesign phase, and the desired schedule leading up to ultimate occupation of the facility. All of these elements may not be completely understood prior to project onset, but they must be continuously monitored and adjusted throughout the evolution of the design. Any clarification of previously unknown information will expedite the process and improve the efficiency of initial efforts to advance the design. With much of this criteria identified, the institution can then select a team of design professionals to assist with continued development of the project. During the design phases, the team should clarify initial assumptions through continuous information gathering, and establish and document key performance indicators for the project.

The programming and predesign phase is perhaps the most critical, and yet the most often overlooked, phase of a laboratory design project. As mentioned earlier, there are often significant pressures to begin design or construction without allowing for adequate predesign thinking and activities, however, "only after a thorough search for pertinent information can the client's design problem be started" (Peña and Parshall 2012). In the programming and predesign phase, all of the project information, risk assessments, space needs, structure of the organization, project goals, critical relationships, equipment needs, scientific capacity, growth assumptions, regulations to be met, and any other ideas brought forward need to be identified and organized in a manner that will help the design team incorporate all this information into a well-considered design. Prior to commencing with design, this information should be organized into the following components:

- Project goals, aspirations, and requirements, including risk assessments
- Local or national codes, regulations, guidelines, and standards to which the project will adhere. (The International Building Code (2012), though not required in all areas, may serve as a good addition in helping to ensure safety and security.)
- A functional space program—listing all program spaces required in the facility
- Organizational charts—illustrating the organizational structure of the group(s) that will use and operate the facility
- Relationship diagrams—illustrating critical relationships between spaces and user groups
- A listing of institutional space standards (if available)
- Room diagrams and room data sheets—recording as much information about each room type, its equipment, and service needs as is available

INFORMATION GATHERING

When laboratories have existing work spaces, designers should tour through these areas and ask users to express what elements of their current space work well for them, and what aspects they would adjust to improve functionality and spatial efficiency. However, it is equally important to gather information related to the risk assessment, such as the types of scientific processes carried out in each space, the agents under study, the equipment used for this work, the number of persons involved in the work, and the relationships to other areas. In addition to gathering information about the types of spaces required and the working methods to be accommodated in new laboratories, the design team must obtain any other necessary information (e.g., training programs, hours of operation, security policies) to support the biorisk assessment and to develop a consensus about the safety and security needs of the biological agents or toxins under study in the facility.

Scientific working methods constantly evolve, whereas physical facilities are generally static. Equipment can be changed, although facility design will often constrain the capacity for change. Facilities can be renovated, but there are limitations to what can feasibly be done to an older building, and often funds for changing spaces are not available. Inevitably, over time, scientific working methods will be adapted to a less than ideal space. Therefore, the design or programming team should identify where new types of space may be required in order to provide a new facility that is better suited to the current best scientific practices. It can also be invaluable to study, and relate to users, the methods by which other, newer facilities carry out similar work. Benchmarking tours and discussions with peers from other institutions can help users to think about the ideal type of space for their work.

Space standards and care requirements for animals also evolve over time. Where animal use is required for research and diagnostic purposes, the design team should review local practices for animal care and also consult the "Guide for the Care and Use of Laboratory Animals" (National Research Council 2011).

SCHEMATIC DESIGN

In the schematic design phase, the design team should progress through an iterative design process, developing options and reviewing these with the appropriate project stakeholders. It is critical at this stage to ensure that all necessary space adjacencies and configurations are correctly established. The schematic design of a facility provides the framework for implementation of all design-based, and many protocol-based, risk mitigation measures. The decisions made when developing the schematic design for a laboratory facility will eventually have a profound influence on the design of biorisk management systems and protocols; they will influence the functional relationships and the way the building is built, the way it is serviced, and the way it is secured. Subsequent sections in this chapter discuss design best practices that should be considered during the schematic design and design development phases of the project.

LABORATORY DESIGN BEST PRACTICES

PUBLIC-PRIVATE SEPARATION

The separation of public and private space is a principle inherent in the planning of most any type of building, and this separation is generally a gradual shift, rather than a hard division. For example, a residential house may have a yard that is open (or at least visible) to all. Upon entering the house an individual will first come upon the public spaces, such as the living and dining areas where guests are invited in; beyond this, one may encounter the kitchen, where perhaps only close friends and family may gather, and further into the house can be found the private sleeping areas. The separations between these spaces are primarily established by their location within the building, and by walls and doors that distinguish one space from another. In a laboratory facility, it is important to establish similar separations between public and private areas.

In most laboratory facilities, there will be areas to which the public has free access, and areas where public access is monitored, controlled, or perhaps completely restricted. Often, the secure areas begin at the boundary of the site. Once inside the site, pedestrian movement may be monitored and, to some degree, controlled by roads and walkways. At the entry to a laboratory building, a visitor may encounter a locked door or may be freely allowed into a small entryway, but may require some type of permission or escort in order to proceed further into the facility. Once a member of the public or a visitor has access inside the building, there may still be some public space, where visitors may freely move about. This area may consist of a lobby, restrooms, and conference areas. The offices, laboratories, and supporting areas beyond, however, will be defined, to a greater or lesser degree, depending on the nature of the facility, as private space. Access to these private spaces should be monitored and controlled with security devices in a manner appropriate to the level of risk associated with the work that takes place there. The proper organization of public and private spaces within a laboratory facility forms the beginning of what can be referred to as a zoning strategy.

ZONE STRATEGIES

While the division between public and private space in a laboratory facility is a critical design driver, it is also important to look at the organization of different types of spaces or zones of space within the building. The zone strategy (sometimes referred to as the big picture organization) of a laboratory facility will have a profound influence on the design of biosafety systems and protocols. It will influence the functional relationships, and the way the building is built, serviced, and secured.

When studying the overall zoning of the facility, the design team may decide to put animal holding rooms in close proximity to the loading area, to save animal handlers time and energy for years to come. Or there may be a decision to put all containment areas, spaces that contain high-risk work, together toward the center of the building so they are well isolated, and the security systems can be concentrated on a single area. Or there may be a decision to put all the spaces that need robust construction and intensive services all in one wing, and all the lighter construction in another wing, consolidating the more expensive areas to save on overall costs. Figure 4.3 illustrates three potential zoning diagrams for a common group of space types.

It is important to understand the zoning strategy of a facility, even in an existing building that already has the primary relationships established. By understanding the overall organization, designers and decision makers can be better prepared to make renovations to the building, to solve problems, or to make operational changes that work well with the given context.

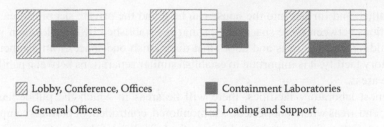

☒ Lobby, Conference, Offices   ■ Containment Laboratories

☐ General Offices   ▤ Loading and Support

FIGURE 4.3 Zoning concept diagrams.

FLOW ANALYSIS

Flow analysis is an important activity that should be repeated throughout all phases of the design process. Typical flows that should be addressed and understood include material movement through the building, sample delivery and distribution, personnel circulation, paths of travel for waste with appropriate identification of key inactivation destinations, as well as short-, medium-, and long-term storage areas for chemical, biological, radiological, and nonhazardous waste. Animal facilities have additional flow considerations that should be understood early in the design process. Food, bedding, clean and dirty paths of travel for cages and racks, quarantine and isolation areas, as well as carcass handling considerations should be carefully analyzed.

In the early stages of design, the flows should help to form the basis for the overall zoning strategy that will shape the building. Critical flows should be mapped out in an abstract fashion even prior to embarking upon the design process to ensure that all parties have a common understanding of the relationships that need to be supported by the facility design.

The diagram shown in Figure 4.4 illustrates the flow of samples through a new facility prior to developing a floor plan. It is also important to analyze flows, in a progressively more detailed fashion as the design progresses. Analyzing flows can help designers choose between alternate concepts during the schematic design phases. Illustrating the paths that users, materials, and waste will take through the floor plans will help to determine the relative efficiency of one layout over another. In the late stages of design, flows should be studied in detail by mapping out step-by-step protocols in critical areas. This process will help to ensure that all the necessary space, equipment, casework, storage areas, and services are in the correct place to

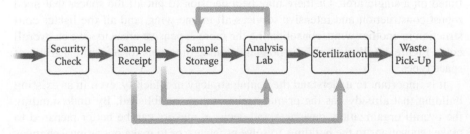

FIGURE 4.4 Sample flow diagram.

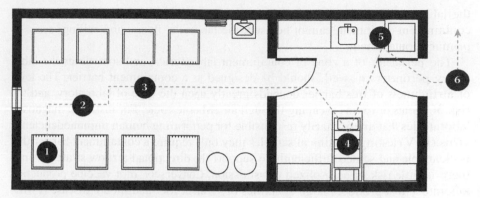

1. Animal subject sedated in cage
2. Subject removed once asleep
3. Subject placed on cart, covered, and moved to procedure
4. Subject placed on table, blood or tissue sample taken
5. Sample packaged at bench and placed in double container
6. Sample taken from procedure area to lab for analysis

FIGURE 4.5 Protocol mapping diagram.

support the work. The diagram shown in Figure 4.5 illustrates the process of taking a sample from a research subject for study in a neighboring laboratory. Looking at the details of the sample transfer process will help to identify practical needs, such as the cabinetry required to store transfer containers, and may also uncover risks not previously considered. For instance, it may be discovered at this point that users leaving the lab to transfer the samples pose a risk of contaminating the corridor. In order to mitigate this risk, the institution may decide to develop a protocol whereby all samples are taken out at the end of the day when users remove contaminated gloves and PPE and prepare to leave, or it may decide to add a passbox to allow samples to be transferred to personnel who work in the corridor without the need to open the procedure room door.

LAYERS OF PROTECTION

The concept of a box within a box, which adds layers of protection to any laboratory design project, uses barrier separation to mitigate the potential accidental release of dangerous materials. When the risk of infection by exposure to an infectious aerosol is present, higher levels of primary containment and multiple secondary barriers may become necessary to prevent infectious agents from escaping into the environment (US Department of Health and Human Services 2009). Within any containment laboratory, the best practice is always to keep infectious or dangerous materials within some type of primary containment device. In bioscience laboratories, the most common of these is the biosafety cabinet or BSC. Glove boxes and animal isolation cages are other means of providing primary containment. With equipment providing the primary means of containment, the laboratory space then provides secondary containment, preventing anything that has escaped primary containment from escaping the lab. In some cases, such as when working with large animals, where primary containment equipment cannot be used, the lab walls, floors, and ceilings serve as primary containment.

The perimeter of a zone of containment laboratories, and sometimes the laboratory perimeters as well, should be designed as a containment barrier. The level of airtightness of this barrier depends greatly upon the type of laboratory, and the risk of agents or toxins escaping through an airborne route. For instance, diagnostic laboratories that are primarily responsible for performing human immunodeficiency virus (HIV) testing on clinical samples may only require a containment barrier that is cleanable and sealed sufficiently to support the directional airflow strategy, since there is little risk of aerosolized transmission. Conversely, in a vaccine production laboratory that produces large quantities of virulent foot-and-mouth disease (FMD) virus, there is the potential for large spills, capable of generating significant aerosols. Therefore, a more robust barrier is required to support more stringent requirements for directional airflow and gaseous decontamination. Often the level of airtightness required of a containment barrier will depend as much upon the intended method of decontamination as upon the risk of aerosols being generated in the lab. Laboratories that will be decontaminated with gas should have containment barriers tight enough to contain that gas in a static or even slightly pressurized state (depending upon the method of decontamination anticipated) so that the decontaminating agent does not become a risk to users or damage areas of the facility not designed to withstand exposure to strong decontaminating agents. Agreement (among the design team, safety personnel, operations and maintenance personnel, and those responsible for constructing the barriers) on the level of airtightness required for containment barriers, and where those barriers will be defined, is critical to designing laboratory spaces specific to the risks. Drawings at all stages through the development of the design should illustrate where and what type of containment barriers are planned so the spaces, systems, services, equipment, and personnel protocols for crossing the barriers are developed as part of that barrier. Clear definition of containment barriers will help the design and engineering team to develop a safe facility, and will help the building occupants to use the facility safely.

The box within a box strategy can also add to the overall security strategy by incorporating facility controls and limiting access for unauthorized personnel into progressively more secure areas of the building (Salerno and Gaudioso 2007). Access to the biological agents and toxins is usually limited by the institution to only those individuals who require access as part of their job responsibilities. For example, these materials may be stored within a secured freezer or cabinet, the freezer or cabinet within a lockable room, the room within a secure zone, within a secure facility, and upon a secure site. This layering concept, as illustrated in Figure 4.6, is appropriate for all types of laboratory facilities. Whenever an individual crosses the boundary between a less secure zone and a more secure zone, there should be some type of access control that prevents unauthorized personnel from gaining inappropriate access. The measures and devices used for controlling access will vary greatly, and may even include devices that have the ability to monitor when a specific individual entered or exited a facility or a specific laboratory. Monitoring and control may be as simple as a security person, who monitors entries and exits, or more complex, such

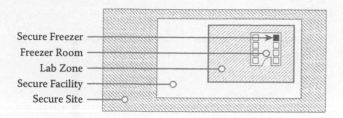

Secure Freezer
Freezer Room
Lab Zone
Secure Facility
Secure Site

FIGURE 4.6 Security layering concept.

as biometric devices (e.g., iris scan, hand geometry, fingerprint) and card readers that record exact times of personnel access.

The security measures and types of devices used should be appropriate to the risk, and also appropriate to the budget and operational capabilities of the institution (Salerno and Gaudioso 2007). For instance, access to a diagnostics facility responsible for testing clinical samples for common diseases may be relatively open, with free access to public spaces and perhaps a guard who monitors those coming and going. Conversely, access to the site of a research facility that works with highly infectious diseases may be restricted to only authorized personnel. The facility may be secured with perimeter fencing, gates that require card access, and video monitoring of all entries; the facility may require all visitors to be escorted at all times. From a design perspective, the most important aspect is to develop a facility that will support a multilayered security concept, allowing the details of access control to be attuned to, and commensurate with, the risk. In addition, the design should allow for access controls to be modified as needed over time without requiring significant changes to the facility planning.

PLACEMENT OF PRIMARY CONTAINMENT DEVICES

Placement of primary containment devices is a simple, yet critical, aspect in the design of any laboratory. Primary containment devices may include biological safety cabinets (BSCs), chemical fume hoods, glove box isolators, downdraft tables, backdraft tables, or any other piece of equipment that provides a primary protective barrier for staff who work with potentially hazardous materials. While designed to enhance safety for laboratory workers and the surrounding environment, these primary containment devices can have a contradictory impact if improperly placed.

When a risk assessment indicates that part or all of the lab users' work requires a primary containment device, such as a BSC, in order to protect the users or the environment, the users rely on the airflow patterns within and at the open face of the cabinet to prevent aerosols from escaping into the lab. Disruption of these airflow patterns can compromise the effectiveness of a BSC; therefore, the cabinet should be located such that a person can perform his or her work without other laboratory personnel passing by and disrupting the airflow of the cabinet. When the risk assessment indicates that the work presents a relatively low level of risk, BSCs may be located at the end of the lab, and shared equipment and sinks at the front of the lab. This arrangement is usually sufficient to minimize the risk of one user disrupting

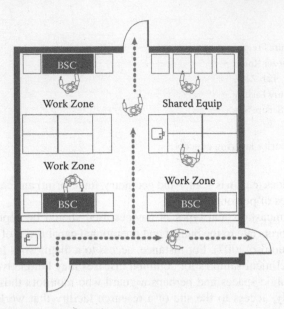

FIGURE 4.7 Placement of primary containment devices.

the airflow, and potentially compromising containment of another user's cabinet. When staff need to move through a lab frequently, or in a lab where highly infectious agents are studied and the consequences of disrupting cabinet airflow are very high, the circulation should be arranged so that the working areas in front of the BSCs are isolated and well protected from disruption. Figure 4.7 illustrates a lab layout with several BSCs, each having a separate working zone located away from circulation. If the risk assessment determines a high risk of aerosol transmission, the BSC, supporting bench space, and equipment should be located in a separate room.

FACILITY DESIGN FACTORS

SUSTAINABILITY

Sustainability is a term currently used to describe a multitude of environment-friendly concepts in relation to facility design. In some circles, sustainability is considered to be the successful incorporation of green building technologies in a design project. "The green building movement strives to create a permanent shift in prevailing design, planning, construction, and operational practices toward lower impact, more sustainable, and ultimately regenerative built environments" (US Green Building Council 2012). In other circumstances, *sustainability* is defined as the ability to maintain a facility for the maximum amount of time, using less energy, without compromising safety and functionality. "Energy-efficient laboratories offer the research community cost savings and safer working conditions in addition to serving the larger social good of reducing energy consumption" (Applications Team 2008). From a risk-based design perspective, *sustainability* can be defined as the incorporation of appropriate design solutions that meet the needs of an institution, provide personnel and environmental protection, and are within the institution's capacity for long-term operations and maintenance.

In order to be sustainable, laboratories should be designed to allow construction with locally available materials (locally made or routinely imported) and construction methods, provided the available products meet the minimum quality standards required for the laboratory design. By using locally available materials, methods, and equipment wherever possible, the resulting facility will be less expensive to maintain and will be more likely to be kept in good running order because skilled service personnel and materials for repairs and changes will be readily available. To be sustainable, the facility also needs to be built with operating costs and resources well understood and appreciated by all parties. If a facility is built within the institution's ability to provide staff and material resources to keep the building(s) operating as intended from both a laboratory design perspective and an efficient, maintainable building perspective, then the facility will be truly sustainable. An effective operations and maintenance program should conserve energy and water and be resource efficient, while meeting the comfort, health, and safety requirements of the building occupants (National Institute of Building Sciences 2014). A sustainable facility, which is well supported, well maintained, and well staffed, is less likely to be subject to changes or cost-cutting measures that could compromise security or safety and potentially increase risk at the facility.

ADAPTABILITY AND FLEXIBILITY

Most design and construction projects are expected to maintain a useful life of 30 years or more before they are replaced by new facilities. However, changes in science, mission, instrumentation, and test methodologies pose significant challenges when trying to adapt aging facilities to current trends. For this reason, it is important to prioritize design solutions that allow for ease of future adaptation without significant capital investment and extended downtime for renovations. Wherever possible, the design team and user groups should work together to develop spaces that are both flexible and adaptable.

A flexible space may be used for multiple purposes without significant changes. An adaptable space can be easily modified to suit a new purpose. In the example shown in Figure 4.8, the holding and procedure areas are flexible and can be used for multiple species (in cages) simply by changing the equipment. The area could also be equipped with an extra floor drain and extra mechanical penetrations so that a shower could be added without major changes to the supporting systems to make the space easily adaptable for use with noncaged animals (e.g., sheep).

Flexible design can be very beneficial to lab safety and security. If a flexible or adaptable lab space allows a new function to be located within an already established containment or security zone, then the new function will have little impact on the established safety and security operations. If a new function cannot be accommodated and has to be placed in a disconnected area, then established security and containment boundaries will be affected. This could lead to secure materials being moved through nonsecure areas, increasing risk, and inefficient work patterns if people and materials have to move in and out of different secure zones.

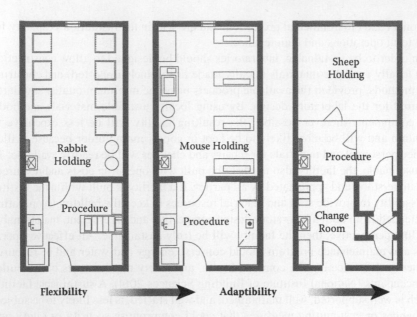

FIGURE 4.8 Flexibility and adaptability of same room spaces.

CONCLUSION

A risk-based design strategy should place a strong emphasis on identifying the specific mitigation measures needed to reduce the risks associated with handling and storing infectious biological materials, rather than implementing mitigation measures solely from a biosafety level or predefined solution perspective. This strategy will decrease the likelihood of creating a situation in which a facility includes overcomplicated engineering elements or unsustainable systems. Biological safety measures are put in place to prevent accidental exposure to biological agents, and to prevent the unintentional release of agents into the surrounding community and the environment. Typically, this is achieved through a combination of design features, specific biosafety mitigation measures, and biorisk management SOPs. Biorisk mitigating measures can be most efficient and effective when all parties involved in the operation of the facility understand how the facility design and the protocols for using it will work together to promote safe and secure operations. By bringing together a wide-ranging group of stakeholders, and developing a common understanding of the risks associated with the work at hand, a design team can develop a facility that is well suited to mitigate the biorisks present and is sustainable within the community, the budget, and the capabilities of the operating institution.

The correct design of a facility can help prevent the accidental release of biological agents into nonlab areas, and possibly into the environment, by incorporating features such as containment barriers, HEPA filters, and directional airflow. It is equally important to include features that help promote compliance with the established SOPs. For instance, locating a sink near an exit where users need to wash their hands, providing ample space for storage to reduce lab clutter, and providing enough space to promote the proper function of biosafety cabinets will facilitate and encourage users to follow the proper procedures. Ensuring that international biorisk management best practices are being considered throughout the design process will also help lab users to execute their work activities in a safe and secure manner.

Biological security measures are intended to prevent the theft or intentional release of biological agents or toxins by insider or outsider threats. Facility design can support biosecurity efforts by ensuring that access to the laboratory zones can be adequately restricted and monitored, secure rooms can safely store dangerous agents, and laboratory infrastructure is appropriately designed and in place to support the security equipment. An example would be the addition of access control devices to secure lab spaces and monitor secure freezers.

KEY CONCEPTS

- The implementation of containment concepts and features in a biological laboratory facility should be based upon a well-considered analysis of risks—not upon predefined, prescriptive standards.
- Biorisks should be mitigated with a complementary combination of protocol-based and design-based measures that specifically mitigate the identified risks for specific facilities. Overdesigning laboratories can complicate operations and make a facility unsustainable by inflating construction and operating costs and installing difficult-to-maintain equipment.
- Design input and information should be gathered from a wide-ranging group of stakeholders, including peers who do similar work in other institutions/ countries, in order to provide the design team with a well-rounded understanding of all risks involved in the work, and a broad-ranging intellectual resource to assist in the development of risk mitigation strategies.
- The zoning of the facility, directing flows of materials and personnel through the facility, and containment barriers within the facility are critical elements of risk mitigation strategies, and should be analyzed and illustrated through all phases of design.
- Biosecurity should be addressed in a layered approach that will allow for the implementation of access control and monitoring measures at multiple points within the facility.
- Facilities that are flexible, adaptable, and sustainable will be more able to preserve established safety and security protocols by allowing for changes in scientific working methods, or even the scientific mission, without the need for major changes to facility infrastructure or major changes to the risk mitigation concepts incorporated into the facility design.

REFERENCES

Applications Team. 2008. A Design Guide for Energy Efficient Research Laboratories. http:// ateam.lbl.gov/Design-Guide/ (last updated September 5, 2008).
European Committee for Standardization. 2011. CEN Workshop Agreement CWA 15793: Laboratory Biorisk Management. ftp://ftp.cenorm.be/CEN/Sectors/TCandWorkshops/ Workshops/CWA15793_September2011.pdf.

International Code Council. 2012. International Building Code. http://publicecodes.cyberregs. com/icod/ibc/2012/index.htm.

National Institute of Building Sciences. 2014. Whole Building Design Guide. http://www. wbdg.org/ (accessed July 28, 2014).

National Research Council. 2011. *Guide for the Care and Use of Laboratory Animals*. 8th ed. Washington, DC: Institute for Laboratory Animal Research, National Academies Press.

Peña, William M., and S.A. Parshall. 2012. *Problem Seeking: An Architectural Programming Primer*. 5th ed. New York: John Wiley & Sons.

Salerno, Reynolds M., and J.M. Gaudioso. 2007. *Laboratory Biosecurity Handbook*. Boca Raton, FL: CRC Press.

US Department of Health and Human Services. 2009. *Biosafety in Microbiological and Biomedical Laboratories (BMBL)*. 5th ed. US Department of Health and Human Services, Public Health Service, Centers for Disease Control and Prevention, National Institutes of Health. http://www.cdc.gov/biosafety/publications/bmbl5/.

US Green Building Council. 2012. *Green Building and LEED Core Concepts Guide*. 1st ed.

World Health Organization. 2004. *Laboratory Biosafety Manual*. 3rd ed. http://www.who. int/csr/resources/publications/biosafety/WHO_CDS_CSR_LYO_2004_11/en/.

5 Rethinking Mitigation Measures

Jennifer Gaudioso, Susan Boggs,
Natasha K. Griffith, Hazem Haddad,
Laura Jones, Ephy Khaemba, Sergio Miguel,
and Cecelia V. Williams

ABSTRACT

Once a facility is constructed, risk managers have several tools to mitigate biorisks identified in the risk assessment. They can decide to eliminate the risk or substitute it; they can use equipment to mitigate risks, such as biosafety cabinets, badge readers, or personal protective equipment (PPE); they can decide who will have access and execute the work; and they can change work practices and administrative controls. Although these are all elements that should be considered at the time of design of a facility, they are also the same tools available to adjust the mitigations as needed to accommodate changes in mission. Risk managers must understand the various options that can be used to mitigate risks because no matter how well planned a facility is, the mission will inevitably evolve and risk mitigations will need to be reevaluated.

INTRODUCTION

Current Western occupational hygiene literature recognizes a hierarchy of mitigation controls as elimination, substitution, engineering controls, administrative controls, practices and procedures, and personal protective equipment, each having various advantages and disadvantages (DiNardi 1997; NIOSH 2010; OSHA 2014). Engineering controls are subdivided into primary and secondary controls: primary controls are safety and security equipment, while secondary controls refer to the facility (see Chapter 4). Primary engineering controls include the equipment used in the laboratory to protect laboratory personnel and prevent accidental release, or intentional removal of biohazardous materials from the laboratory. Examples of this equipment include biosafety cabinets (BSCs), chemical fume hoods, access controls (e.g., keys, cipher locks, badge swipes, and biometric readers), alarms (e.g., fire alarms, low oxygen sensors, motion sensors, and door open alarms), and other specialized equipment. Administrative controls can include policies, such as decisions about which personnel will conduct work, and training. Practices and procedures codify the expected behaviors of personnel. For example, expectations for waste handling should be captured in a documented procedure. Another standard practice in a bioscience facility is the use of mechanical pipettors instead of mouth pipetting. Personal protective equipment (PPE) is equipment worn by personnel that is designed to reduce their exposure and protect them from injury. Common PPE in bioscience facilities includes goggles, gloves, lab coats, and respirators.

Gressel (2005) advocates that elimination and substitution merit particular attention in this hierarchy because these options not only increase the level of protection to the worker and the work environment, but also may result in mitigation approaches that are less expensive and require less maintenance. Similarly, Soule (2001) explains how elimination and substitution frequently offer the most effective solution to an industrial hygiene problem. From a biorisk management perspective, the benefits of elimination or substitution need to be weighed against any scientific impacts. For example, there is an active debate over the benefits of retaining Variola major virus for research. The World Health Organization (WHO) has decreed that research on live Variola major virus must have public health benefits and not simply enhance scientific understanding (Butler 2011). After eradication of smallpox in 1980, WHO and member countries readily agreed to consolidate the remaining virus isolates in two laboratories to eliminate the biosafety and biosecurity risks at all other institutions. For work with the only other eradicated virus, Rinderpest, the World Organisation for Animal Health (OIE) lifted the moratorium on research with live virus and implemented a process for reviewing research proposals against three criteria to determine if the scientific benefits outweigh the biorisks (OIE 2013). These two examples showcase elimination as a risk reduction strategy, but many times substitution of a less pathogenic strain also can substantially reduce the risks while yielding good science. However, it is critical to ensure that the substitution option is actually less risky than the original process. In 2004, the Children's Hospital Oakland Research Institute believed it was working with nonviable vegetative cells of *Bacillus anthracis*-Ames strain as a substitute for the pathogenic material. When it was discovered that the specimen had not actually been inactivated, eight personnel had to receive postexposure chemoprophylaxis for prevention of inhalational anthrax (MMWR 2005).

When elimination or substitution of hazards is not feasible or may not provide comprehensive solutions to the risk, engineering controls are often implemented to reduce the risk. The phrase "hierarchy of controls" leads many to believe that engineering controls are the most important aspect of biorisk management. However, engineering controls are often misused and, as such, can provide a false sense of safety or security. Nevertheless, this is a common misperception, and as a result, some laboratories are designed entirely around engineering controls, neglecting other equally important control elements, such as administrative controls, standard operating procedures (SOPs), and the use of PPE. The level of implemented engineering control should be proportionate to the risk and should work in conjunction with other controls to optimize overall risk mitigation. Furthermore, disproportionate reliance on engineering controls to reduce laboratory risks can lead to overdesign of facilities or exorbitantly high operation, maintenance, and sustainability costs (see Chapter 4). Instead of being the single point of control, engineering controls should be approached as one aspect of a mitigation strategy that blends engineering controls with other elements, such as elimination, substitution, administrative controls (including training and mentoring), SOPs, and PPE. There are many risk factors and risk mitigation strategies that need to be considered when choosing the appropriate mitigation measures, including, but not limited to, agent characteristics, endemicity, population susceptibility, availability of prophylaxis and treatment, availability of trained and experienced personnel, and availability of resources. This is why the risk assessment (Chapter 3) is the crucial first step in selecting situation-specific mitigation measures.

Readers should rely on standard industrial hygiene, biosafety, and biosecurity texts for specific technical details on mitigation measures for bioscience institutions (WHO 2004, 2006; US Department of Health and Human Services 2009; Plog and Quinlan 2012). However, we argue that these cornerstone references are best used as a menu of options for selecting biorisk mitigation measures and not simply a checklist to implement measures based on default biosafety levels. Historically, the design and implementation of mitigation measures have been based upon the biosafety level (BSL) of the laboratory (e.g., BSL1, BSL2, BSL3, or BSL4). The WHO states that "biosafety level designations are based on a composite of the design features, construction, containment facilities, equipment, practices and operational procedures required for working with agents from the various risk groups.... The biosafety level assigned for the specific work to be done is therefore driven by professional judgment based on a risk assessment" (WHO 2004). The use of the AMP model builds on this approach, further enabling professional judgment in identification and implementation of specific mitigation measures based on a thorough risk assessment rather than relying on the predefined solution sets of biosafety levels. Using the biosafety level method to identify mitigation measures to be used is certainly better than no method at all, but a more strategic and technical approach to implement control measures would be to use a situation-specific risk assessment to more effectively allocate limited resources and reduce risks. By using the AMP model to select mitigation measures to address the identified risks, an institute does not necessarily have to use all of the elements in the hierarchy of controls, but rather can rely on assessment and performance to help ensure that risk is reduced to an acceptable level. The effectiveness of mitigation control measures selected must also be evaluated on the feasibility and practicality to implement and sustain the measures.

CASE STUDY: CHALLENGES MITIGATING BIORISKS—TEXAS A&M UNIVERSITY

Although many laboratories successfully implement measures to mitigate their biorisks, the following case study is instructive as a source of lessons learned. On April 20, 2007, Texas A&M University (TAMU) in College Station, Texas, received a cease and desist order from the US Centers for Disease Control and Prevention (CDC) for any and all manipulations and storage of *Brucella abortus*, *Brucella melitensis*, and *Brucella suis* (Kaiser 2007; Weyant 2007). On June 30, 2007, the CDC expanded the cease and desist order to include all work with select agents and toxins while CDC conducted a "comprehensive review" to determine if TAMU met the standards for handling select agents (Schnirring 2007) and delineated specific violations related to lapses in specific mitigation measures. In addition, the principal investigator of the *Brucella* laboratory was suspended.

These unprecedented cease and desist orders stemmed from TAMU's failure to report to CDC two cases of exposure to select agents in 2006. The first exposure occurred in February 2006 to a lab worker who had cleaned a Madison Aerosol Chamber that had contained *Brucella* in a biosafety level 3 laboratory in the School of Veterinary Medicine. The lab worker subsequently developed brucellosis and recovered after treatment with antibiotics. One month later, three other workers from the TAMU Medical School tested positive for antibodies to *Coxiella burnetii*, the bacterium that causes Q fever, but did not develop the illness. TAMU admitted that it had failed to report both incidents in a timely fashion. Five laboratories in the School of Veterinary Medicine and the Medical School with 120 workers were closed. According to the CDC, this was the first time an entire university's select agent program was suspended.

These incidents raise the following questions: How did these events occur at a highly respected and well-funded university? What safeguards were operational? Were biosafety/biosecurity good laboratory work practices, standard operating procedures, administrative controls, personnel management, record keeping, incident response planning, and biorisk management in place?

The CDC report of August 31, 2007, indicated that TAMU had an inadequate biosafety and biosecurity program—violations occurred with primary biosafety and biosecurity equipment, administrative controls (especially with regard to personnel management) and procedures, and personnel management. These violations to the select agent regulations (42 CFR 73) included over 25 institutional violations, as well as over 45 violations attributed to the specific principal investigator's research, laboratories, and employees. For example, safety equipment was not used properly: the Madison Aerosol Chamber used for animal studies opened directly into a research laboratory with no primary containment barriers, clearly highlighting the absence of a system that systematically evaluated the performance of the risk mitigation measures. TAMU was also cited for failing to report a release from containment.

TAMU had difficulty implementing mitigation measures that intersected primary controls and administrative controls. At least seven incidences of unauthorized access to select agents occurred because either the primary engineered access controls did not work properly or the associated administrative controls for personnel management policies regarding who was authorized to have access were missing or not performing properly. Other specific administrative control failures documented by CDC included:

- Failure to obtain approval for select agent work prior to experiments being conducted with both *Brucella* spp. and nine *Coxiella burnetii* aerosolization experiments. This was a clear failure to implement or verify performance of administrative controls for work planning and authorization.
- Individuals with the greatest access to laboratories and animal rooms did not meet proper medical entry requirements. No effective medical surveillance program was in place. This is another administrative controls failure regarding personnel management.

- TAMU's approved certificate of registration did not match the list of individuals provided by the PIs—yet another failure in the administrative controls for personnel and work approvals.
- TAMU lacked training records for individuals with approved access and documentation on formal training programs for individuals who worked in the laboratories of the PIs. Documented performance for training and other administrative controls was insufficient or nonexistent.
- The security plan did not adequately address transfer of select agents or toxins. There was no documentation that a security plan had been designed in accordance with a site-specific risk assessment. Assessment was critically absent from the development of the security administrative controls.

Clearly, TAMU did not use an AMP approach for developing and implementing procedures and practices. CDC cited TAMU for inadequate administrative controls for preventing exposure (SOPs, routine maintenance) and waste handling procedures. CDC also noted inventory discrepancies and deemed institutional inventory oversight for select agents inadequate. Utilizing the assessment and performance components of the AMP model could have helped TAMU develop more appropriate SOPs for these key activities.

Each cited deficiency could have been avoided. Instead, TAMU had to pay $1 million in fines for the violations (Schnirring 2008) so that the university could resume its biodefense research. The large monetary penalty set a new standard of accountability for all research institutions that conduct work on biological select agents. However, the most significant impact was related to the publicity the incidents generated for TAMU, which tarnished the university's reputation and likely negatively impacted TAMU's failed attempt to win approval for a major new federal laboratory: the National Bio and Agro-Defense Facility.

USING AMP TO STRENGTHEN MITIGATIONS

So, how can these types of negative impacts be avoided? And, how can the AMP model be applied to optimize the implementation of the mitigation measures— measures that can be adjusted after a facility is built as the science changes? Institutions must determine what primary engineering controls to employ, which people to perform what activities, what practices and procedures to implement, and what PPE to require. AMP is a simple tool to use to make these decisions, ensuring a holistic, strategic, cost-effective, and sustainable approach.

PRIMARY ENGINEERING CONTROLS

Primary engineering controls are an integral component of biorisk management that can substantially mitigate biorisks, when used in accordance with a comprehensive risk assessment and a solid understanding of how the performance of these controls will be monitored and maintained. There is a critical interplay between the primary engineering controls, the procedures, and the personnel. As an example, we will discuss some of these relationships for the biological safety cabinet (BSC). The BSC

(Kruse et al. 1991) is a common and critical primary engineered control for reducing the risk of cross-contamination (product protection), reducing the risk to the worker of an aerosol or droplet exposure, and reducing the risk to the environment of an aerosol exposure—but only if it is installed, maintained, and used correctly. If there are air drafts from heating or air conditioning, personnel movements, doors opening and closing, or other sources of air drafts, the performance of the BSC will suffer. In fact, if a procedure that creates aerosols is being conducted in a BSC, the amount of aerosolized organisms that escape from the BSC is directly proportional to the velocity of the cross-draft (Rake 1978). Some types of BSCs must be hard ducted to the building exhaust, while others can be installed without any connection into the facility ventilation system. It is important to understand these differences since they impact laboratory procedures, including when the BSC is not in use, handling failure modes, and the ability to work with any chemicals.

The National Sanitation Foundation (NSF)/American National Standards Institute (ANSI) Standard 49 establishes independent performance criteria for BSCs (NSF/ANSI 49). The US *Biosafety in Microbiological and Biomedical Laboratories* (BMBL) (US Department of Health and Human Services 2009) recommends that laboratories certify their BSCs against this standard before being placed into service, after being relocated, and annually to ensure proper functioning of this critical primary control for biohazard risks in most laboratories. However, Kruse et al. (1991) document examples of improperly certified BSCs, and how these primary controls did not perform as intended and inadvertently failed to mitigate the risks in the ways the facilities assumed. In one example, the protective covering of the filter for shipping of the BSC had not been removed, so air was not filtered and exhausted properly. Instead, air blew out of the front of the BSC into the worker, yet the BSC had been tested and certified four times over several years (Kruse et al. 1991). To address this personnel competence failing of the certifiers, NSF started a program to accredit BSC certifiers in 1993 (US Department of Health and Human Services 2009).

The level of protection depends on the mechanical performance of the primary engineering control device as well as good laboratory work practices (Kruse et al. 1991). If the personnel who use the BSCs do not understand and follow the correct procedures for conducting work inside the BSCs and decontaminating the BSCs afterwards, the BSCs will likely not mitigate the risks properly, even when the BSCs are properly selected, installed, and functioning correctly as verified by certification. Poorly trained workers often use the air intake grill of a BSC as part of the work surface, covering the grill with an absorbent pad, microfuge tube holders, or other equipment in use in the cabinet. These items disrupt the protective airflow. In this case, the worker may assume that certain protection is offered, does not attempt to augment the protection with additional PPE, and performs the procedure. If the worker were aware that the containment aspects of the BSC were hindered, he or she might choose to find an alternate mitigation strategy or choose not to perform the procedure. There are many other best practices for working in a BSC that a worker must be willing to follow if the risks are to be mitigated as planned even if, in doing so, additional time is required.

STANDARD OPERATING PROCEDURES

Despite the plethora of engineered controls available to a bioscience institution, the success of these controls depends primarily on individual workers using the controls as designed. Standard operating procedures (SOPs) are the primary tool to achieve this outcome. These instructional documents are designed to guide "different people doing one thing the same way and achieving the same outcome" (Kaufman 2009). SOPs generally aim to achieve a single or small outcome (e.g., how to correctly wash hands). Examples of SOPs one might expect to see in a bioscience laboratory include, but are not limited to: (1) entering/exiting laboratory, (2) donning/doffing PPE, (3) instrument operating procedures (PCR, centrifuge, autoclave, etc.), (4) use of biosafety cabinets, (5) emergency response, (6) hand washing, (7) waste segregation, management, and disposal, (8) inventory control, and (9) experiment-specific activities. These SOPs should be based upon a robust risk assessment of the activities being conducted, the biological agent(s) involved, and the specific primary and secondary engineering controls that are in place for the given facility.

The BMBL and WHO *Laboratory Biosafety Manual* (LBM) list specific practices and procedures by biosafety level (US Department of Health and Human Services 2009; WHO 2004). Yet, practices and procedures are the mitigation measures that can be the most responsive to changing risks; thus, default practices and procedures tied to biosafety levels should not automatically be used. In 2004, while severe acute respiratory syndrome (SARS) virus was still quite new and had not appeared in Belgium, Herman and colleagues (2004) analyzed the laboratory-acquired infections in Singapore and China to inform a risk assessment for different diagnostic protocols. They then used these data to guide the establishment of SOPs, including work with inactivated clinical specimens, because such specimens might still contain infectious RNA, and for storage of positive clinical samples. They also developed other risk-based recommendations for a series of other practices and procedures for handling SARS virus.

Practices and procedures should be accessible to all relevant laboratory staff, and these must be evaluated and validated to ensure that individuals understand and can physically accomplish the procedure. As with other elements of the biorisk management system, the performance of all practices and procedures should be reviewed regularly and when changes occur. To consistently measure the ongoing effectiveness of a practice or procedure, systematic observation of behaviors by coworkers and biorisk management officers can be used in addition to self-reporting or reporting by coworkers.

Gidley Amare (2012) argues that SOPs are fundamental elements of an effective management system that "help cultivate transparent functions, implement error prevention measures and facilitate corrective actions, and transfer knowledge and skill." Although practices and procedures should define how personnel actions fit into the biorisk management framework, persuading individuals to implement standard practices and procedures can be challenging. Amare (2012) explains how some personnel feel that standardization of procedures and practices "diminishes their importance at work and so are unwilling to share their knowledge and skills.... Some workers feel insecure in their position if everybody knows their skills and knowledge." The potential perceived impact of SOPs on job status and job security highlights the importance of the people—the scientists, technicians, administrators, support staff, and others—in the biorisk management system.

PERSONNEL

How should management encourage appropriate behavior among the staff toward biorisk management? How should management monitor personnel reliability? Personnel management throughout the life cycle of the employee is often disconnected from the biorisk management program, yet it should be an integral part of the system. Institutions need to recruit the appropriate individuals who have the necessary technical skills and right attitude, but also need to create an environment where the staff members embrace biorisk management. Every member of the workforce should support the biorisk management vision demonstrated and communicated by the institution, including management, biorisk management advisors, and principal investigators. Universal support of this vision can and will influence adoption of biorisk management practices. But, Burman and Evans (2008) argue that fundamentally, leadership is the key to affecting a safety culture. From the authors' personal experience, when a director attends a biorisk management training course with the workforce, instead of just mandating it for subordinates, he or she demonstrates leadership's commitment and vision better than a memo could ever communicate. The UK Health and Safety Executive identified five indicators of safety culture from the investigation of rail accidents (Human Engineering 2005), including leadership, two-way communication, employee involvement, learning culture, and attitude toward blame. We believe these same factors are fundamental elements in creating a resilient biorisk management culture.

If an institution is successful in creating an impactful biorisk management culture, employees will not feel threatened by the institute's administrative controls, such as SOPs, will accept the need for and not circumvent the engineered controls, and will understand the purpose for not conducting work before receiving authorization. Institutional management needs to assess positions to define the reliability and skills needed, and the subsequent recruitment practices should be commensurate with that assessment and level of risk. Institutions must make decisions about new and current employees' reliability for the position. This can include evaluating trustworthiness, physical competence, mental competence, emotional stability, financial stability, and the ability to uphold obligations to safety, public health, national security, and scientific integrity.

Once an individual is hired, the risk-based approach to personnel management must extend to training (see Chapter 6), support, and career development. The American Association for the Advancement of Science (AAAS) published a report (Berger and Roderick 2014) that discusses strategies for mitigating personnel security risks that touch on all aspects of the employee life cycle, such as hiring, access, employee behaviors, training, personnel actions, and visitors. The report encourages bioscience institutions to rely on performance goals for employees to encourage ownership and a sense of individual responsibility and other mechanisms to build trust and transparency, in addition to more traditional background screening methods and employee assistance programs. In this report, AAAS articulates the elements of personnel security as adherence to security protocols, technical competence, adherence to safety, scientific responsibility, and occupational health and well-being. When human behaviors depart from these norms, personnel can pose a safety or security concern from either malice or disregard (Greitzer and Ferryman 2013). In most cases of betrayal or attack by an employee, that employee exhibited serious personnel problems in the preceding months or year; thus, proactive action to address the anxiety or stress may have prevented the incident (Shaw and Fischer 2005). Additionally, missteps in an employee's scientific responsibilities can negatively impact an institute's reputation and funding. They may also be indicators of the potential for additional misconduct that could lead to safety or security problems. In one of the most comprehensive analyses of scientific misconduct, Daniele Fanelli (2009) determined that "on average 2% of scientists admit to have falsified research at least once and up to 34% admit other questionable research practices." Individuals with admittedly questionable practices in research may disregard the biorisk management practices and pose a risk to the institution and others.

The challenges of personnel, engineering controls, and procedures also converge for visitors. Whether it is the certifier for the BSC or an employee's family member, an institution must assess the risks, develop specific mitigation measures (typically procedural), and validate those measures before admitting any visitor into the institution. The maintenance of laboratory equipment may require visiting technicians to enter the laboratory. Granting access to these technicians may increase the likelihood for theft of material, and also increase the biosafety risk to the individual or environment. Institutions should establish a process to verify the visitor's credentials, ensure material is secured, escort visitors so they are monitored, and decontaminate the laboratory or equipment to be serviced. Equally important is requiring visitors to check out when they leave for the day to ensure accountability for all persons within the facility. For any persons who may require extended access to perform work, additional controls should be enforced, as with employees, including verification of the person's knowledge, skills, and abilities, and employment and education history.

ELIMINATING SAFETY AND SECURITY CONFLICTS

Verifying performance of the system used to mitigate the identified risks will also ensure that conflicts between biosafety and biosecurity are resolved. Do primary engineered controls for security interfere with life safety? Security bars on windows may eliminate an emergency exit route if they do not have emergency release devices installed that allow the bars to be opened from the inside. Personnel also need to be aware of and understand how to use the release devices. Do the access controls operate correctly under the relevant procedures? Primary engineered controls for access can include lock and key, badge swipe, fingerprint reader, or retinal scanner, among others. However, a worker who is wearing gloves cannot use a fingerprint scanner. A physical key or badge may need to be decontaminated if these items are used in a setting where they could become contaminated. Goggles or face shields can interfere with some types of eye scanners. In these cases, the point of access control could be moved, depending on the facility layout and workflows, or a different type of access control equipment could be utilized. It is crucial to consider and balance both biosafety and biosecurity aspects when making decisions about how to mitigate the identified risks. Furthermore, appropriate mitigation measures need to be based on what the infrastructure can support and sustain. For example, personnel can get trapped in the laboratory if the power goes out and there is not an alternate mechanism to open the door or reliable uninterruptible power supply for the locking mechanism.

CASE STUDY: DIFFERENT SOLUTION PATHS TO WORKING WITH EBOLA VIRUS

Since its discovery in 1976 as the causative agent of an outbreak in what is now the Democratic Republic of the Congo, Ebola virus has been designated a risk group 4 agent by WHO, the European Centers for Disease Control, and the US CDC, among others. As a result of this designation, researchers traditionally only handle Ebola virus in a biosafety level 4 laboratory. However, an outbreak that began in Guinea in December 2013 mushroomed into the largest outbreak of Ebola virus disease to date, with active transmission in Guinea, Liberia, and Sierra Leone (as of October 2014). Travelers imported isolated cases into Mali, Nigeria, Senegal, Spain, and the United States (as of October 2014). The magnitude of the outbreak coupled with concerns over the possibility of additional exported cases led several leading public health agencies to release updated guidance for handling specimens suspected of containing Ebola virus to provide recommendations to nonbiosafety level 4 laboratories to safely handle Ebola virus (WHO 2014; PHAC 2014; US Centers for Disease Control and Prevention 2014a).

The new guidelines have many commonalities that focus on implementing specific mitigation measures to match specific facets of the risks associated with handling Ebola virus samples. These guidelines all focus on mitigating the risks of exposure and emphasize the need for risk assessments to identify all possible sources of sprays, droplets, and splashes. The CDC interim guidelines suggest laboratory staff test specimens in a "certified class II Biosafety cabinet or Plexiglass splash guard with PPE to protect skin and mucous membranes" (US Centers for Disease Control and Prevention 2014a). This recommendation combines primary controls to contain droplets created during laboratory procedures with the usage of PPE to mitigate the risks of splashes and other releases from the primary controls. They highlight the risk associated with having laboratory staff work in unfamiliar PPE, which could inadvertently result in exposure during doffing (US Centers for Disease Control and Prevention 2014b). Personnel must be evaluated for their ability and comfort level in executing new protocols to handle Ebola virus. The Public Health Agency of Canada suggests designating specific personnel for work with suspected samples and limiting access to those individuals only (Public Health Agency of Canada, 2014). Notably, the new guidelines for working with Ebola virus do not instruct laboratories to physically change their facilities, but rather to review and adjust their primary controls, administrative controls, personnel, and PPE to handle the potential new risk of a suspected Ebola virus sample.

In a set of questions and answers for "How U.S. Clinical Laboratories Can Safely Manage Specimens from Persons under Investigation for Ebola Virus Disease," the

CDC describes why following protocols for bloodborne pathogens will sufficiently address the risks of clinical labs that handle Ebola virus—even though the CDC itself only works with Ebola virus in a biosafety level 4 laboratory (US Centers for Disease Control and Prevention 2014b). The CDC explains this difference in terms of the risks associated with the different activities conducted since CDC's Ebola researchers grow large quantities of viral stocks for subsequent testing of potential vaccines and treatments, while clinical laboratories primarily process small amounts that are inactivated early in the testing process.

CONCLUSION

In this chapter, we maintain that it is not sufficient for bioscience institutions to simply rely on technical documents such as the US *Biosafety in Microbiological and Biomedical Laboratories* (US Department of Health and Human Services 2009), the WHO *Laboratory Biosafety Manual*, and the WHO *Biorisk Management: Laboratory Biosecurity Guidance* (WHO 2006) for choosing appropriate mitigation measures. To optimize the use of risk mitigation measures, institutions need to embrace flexible, creative thinking about using tools from across the hierarchy of controls to address their specific risks with appropriate biosafety and biosecurity—both in implementing their day-to-day mission and in adapting to disease outbreaks and other mission or situational changes. As the TAMU case illustrates, even a sophisticated institution can encounter serious gaps in its mitigation measures because of a compliance mindset that fails to examine the assessment and performance of chosen mitigation measures. The Ebola outbreak shows how a facility may need to adapt its risk mitigation measures without the luxury of building a new secondary barrier (laboratory). Elimination and substitution of the hazards should be first considerations in any mitigation strategy. In many cases, innovative use of elimination and substitution can also greatly improve the science. The risk of testing for HIV has been significantly reduced through the development of dried blood spot tests that do not need viable virus. At the same time, this advance in technology has improved the ability to conduct HIV surveillance in developing countries since a cold chain is no longer required for the samples (Solomon et al. 2002). However, elimination or substitution may not always be feasible to achieve the scientific mission; the applicability of these control measures needs to be reevaluated regularly as the scientific state of art advances. But, when elimination or substitution is not appropriate or sufficient, facilities can adjust their primary controls, administrative measures, procedures, and PPE to develop multiple strategies to mitigate their biorisks.

REFERENCES

Amare, G. 2012. Reviewing the Values of a Standard Operating Procedure. *Ethiopian Journal of Health Science*, 22(3): 205–208.

Berger, K., and J. Roderick. 2014. Bridging Science and Security for Biological Research: Personnel Security Programs. American Association for Advancement of Science. http://www.aaas.org/report/bridging-science-and-security-biological-research-personnel-security-programs.

Burman, Richard, and Andy Evans. 2008. Target Zero: A Culture of Safety. *Defence Aviation Safety Centre Journal*. pp. 22–27.

Butler, D. 2011. WHO to Decide Fate of Smallpox Stocks. *Nature*. doi: 10.1038/news.2011.288.

DiNardi, Salvatore R. 1997. *The Occupational Environment: Its Evaluation and Control*. Fairfax, VA. American Industrial Hygiene Association Press. pp. 107, 830–831.

Fanelli, D. 2009. How Many Scientists Fabricate and Falsify Research? A Systematic Review and Meta-Analysis of Survey Data. *PLoS ONE*, (5): e5738. doi: 10.1371/journal. pone.0005738.

Greitzer, F.L., and T. Ferryman. 2013. Methods and Metrics for Evaluating Analytic Insider Threat Tools. Presented at IEEE Security and Privacy Workshops.

Gressel, Mike. 2005. Hierarchy of controls and inherently safe design. Proceeding of the American Industrial Hygiene Conference and Exposition. May 21-26. Anaheim, CA.

Herman, P., Y. Verlinden, D. Breyer, E. Van Cleemput, B. Brochier, M. Sneyers, R. Snacken, P. Hermans, P. Kerkhofs, C. Liesnard, B. Rombaut, M. Van Ranst, G. van der Groen, P. Goubau, and W. Moens. 2004. Biosafety Risk Assessment of the Severe Acute Respiratory Syndrome (SARS) Coronavirus and Containment Measures for the Diagnostic and Research Laboratories. *Applied Biosafety*, 9(3): 128–142.

Human Engineering. 2005. *A Review of Safety Culture and Safety Climate Literature for the Development of the Safety Culture Inspection Toolkit*. Research Report 367. Health and Safety Executive.

Kaiser, Jocelyn. 2007. Pathogen Work at Texas A & M Suspended. *Science*/AAAS/News (http://news.sciencemag.org).

Kaufman, S.G. 2009. Evaluating and Validating Laboratory Standard Operating Procedures. Electronic Biosafety Awareness Program, Emory University.

Kruse, R.H., W.H. Puckett, and J.H. Richardson. 1991. Biological Safety Cabinetry. *Clinical Microbiological Reviews*, 4(2): 207–241.

Morbidity and Mortality Weekly Report (MMWR). 2005. Inadvertent Laboratory Exposure to *Bacillus anthracis*—California, 2004. *Morbidity and Mortality Weekly Report*, 54(12): 301–304. http://www.cdc.gov/mmwr/preview/mmwrhtml/mm5412a2.htm.

National Institute for Occupational Safety and Health (NIOSH). 2010. Engineering Controls. Available at http://www.cdc.gov/niosh/topics/engcontrols/.

NSF International and American National Standards Institute (ANSI). 2004. NSF/ANSI Standard 49-2007: Class II (Laminar Flow) Biosafety Cabinetry. Ann Arbor, MI.

Occupational Safety and Health Administration (OSHA). 2014. Hierarchy of Controls. Available at https://www.osha.gov/dte/grant_materials/fy10/sh-20839-10/hierarchy_of_controls.pdf.

Plog, B.A., and P.J. Quinlan, eds. 2012. *Fundamentals of Industrial Hygiene*. National Safety Council.

Public Health Agency of Canada (PHAC). 2014. Interim Biosafety Guidelines for Laboratories Handling Specimens from Patients under Investigation for Ebola Virus Disease. http://www.phac-aspc.gc.ca/id-mi/vhf-fvh/ebola-biosafety-biosecurite-eng.php (updated October 19, 2014).

Rake, B.W. 1978. Influence of Cross Drafts on the Performance of a Biological Safety Cabinet. *Applied Environmental Microbiology*, 36: 278–283.

Schrinning, Lisa. 2007. CDC suspends work at Texas A&M biodefense lab. http://www.cidrap. umn.edu/news-perspective/2007/07/cdc-suspends-work-texas-am-biodefense-lab.

Schrinning, Lisa. 2008. Texas A&M fined $1 million for lab safety lapses. http://www.cidrap. umn.edu/news-perspective/2008/02/texas-am-fined-1-million-lab-safety-lapses.

Shaw, E.D., and L.F. Fischer. 2005. Ten Tales of Betrayal: The Threat to Corporate Infrastructures by Information Technology Insiders Analysis and Observations. Defense Personnel Security Research Center, Monterey, CA.

Solomon, S.S., S. Solomon, I.I. Rodriguez, S.T. McGarvey, G.K. Ganesh, S. P. Thyagarajan, A.P. Mahajan, and K.H. Mayer. 2002. Dried Blood Spots (DBS): A Valuable Tool for HIV Surveillance in Developing/Tropical Countries. *International Journal of STD and AIDS*, 13(1): 25–28.

Soule, Robert D. 2001. Industrial Hygiene Engineering Controls. *Patty's Industrial Hygiene and Toxicology*. New York: John Wiley & Sons. p. 30.

US Centers for Disease Control and Prevention. 2014a. Interim Guidance for Specimen Collection, Transport, Testing, and Submission for Persons under Investigation for Ebola Virus Disease in the United States. http://www.cdc.gov/vhf/ebola/hcp/interim-guidance-specimen-collection-submission-patients-suspected-infection-ebola.html (updated October 20, 2014).

US Centers for Disease Control and Prevention. 2014b. How U.S. Clinical Laboratories Can Safely Manage Specimens from Persons under Investigation for Ebola Virus Disease. http://www.cdc.gov/vhf/ebola/hcp/safe-specimen-management.html (updated October 20, 2014).

US Department of Health and Human Services. 2009. *Biosafety in Microbiological and Biomedical Laboratories*. 5th ed. http://www.cdc.gov/biosafety/publications/bmbl5/.

Weyant, Robbin. 2007. Cease and Desist Order from Violations of the Public Health Security and Bioterrorism Preparedness and Response Act of 2002. Letter to Dr. Richard Ewing, Responsible Official, Texas A&M University. College Station, Texas. April 20. World Health Organization (WHO). 2004. *Laboratory Biosafety Manual*. http://www.who.int/csr/resources/publications/biosafety/WHO_CDS_CSR_LYO_2004_11/en/.

World Health Organization (WHO). 2006. *Biorisk Management: Laboratory Biosecurity Guidance*. http://www.who.int/csr/resources/publications/biosafety/WHO_CDS_EPR_2006_6/en/.

World Health Organization (WHO). 2014. Laboratory Guidance for the Diagnosis of Ebola Virus Disease Interim Recommendations. September 19. http://apps.who.int/iris/bitstream/10665/134009/1/WHO_EVD_GUIDANCE_LAB_14.1_eng.pdf?ua=1.

World Organisation for Animal Health (OIE). 2013. Moratorium on Using Live Rinderpest Virus Lifted for Approved Research. July. http://www.oie.int/en/for-the-media/press-releases/detail/article/moratorium-on-using-live-rinderpest-virus-lifted-for-approved-research-1/.

6 Biorisk Management Training

Lora Grainger and Dinara Turegeldiyeva

ABSTRACT

This chapter will introduce biorisk management (BRM) training in the context of the CEN Workshop Agreement 15793—Laboratory Biorisk Management as a common risk mitigation strategy. The chapter will also address how effective BRM training can be used to address gaps identified in a risk assessment. This section of the chapter will discuss the ADDIE training development model, which provides a framework for identifying training objectives that should always be directly linked to the risk assessment. This section will clearly show the importance of identifying who to train, ways of delivering training, and what to train in order to achieve the desired knowledge, skills, and abilities. The chapter will end with a consideration of the various types of training strategies, including their advantages and disadvantages.

INTRODUCTION

Understanding the risk associated with working with biological materials is fundamental to personal, community, and environmental safety and security. The goal of biorisk management (BRM) training is to equip laboratory managers, administrators, scientists, and other workers with an understanding of the factors that contribute to risk so that risk can be mitigated appropriately. Conventional biosafety training approaches have relied heavily upon memorization of risk categories or levels that depict the risk according to pathogen characteristics or predetermined combinations of mitigation measures. Although these approaches provide an awareness of the risk associated with biological activities, they fall short in providing an opportunity to critically assess the risk. Generally, in the United States, the nomenclature states that laboratories that work with more dangerous agents have a higher biosafety level (such as BSL3 or BSL4) designation (US Department of Health and Human Services 2009). In contrast, in Russia and the former Soviet Union (FSU) countries, microorganisms are classified according to pathogen characteristics, where the most dangerous pathogens are attributed to the lower first class and the least dangerous pathogens to the higher fourth class (Russian State Committee for Sanitation and Epidemiological Oversight 1994). This simple example shows that caution must be taken when assigning numerical nomenclature or applying external risk assessment parameters. More importantly, without fully understanding the situation or other factors that contribute to risk, important risk mitigation decisions may be overlooked.

Risk classifications are valuable sources of information; however, it is clear that they cannot be the only source of information. The active process of performing a risk assessment (Chapter 3), including evaluating pathogen characteristics, potential routes of exposure, and local circumstances and risk mitigation measures, is absolutely necessary because the risk in any situation will vary. Many guidelines are specific to a certain institute, country, or region, and may not apply to every situation. Currently, the nature of life science is changing rapidly with new advances in technology. Therefore, there is a demand for more information to appropriately assess risk in real time.

A common drawback of relying exclusively on risk classifications is that situation-specific risk mitigation measures are overlooked. These measures are often identified following an incident where, unfortunately, hindsight cannot reverse any damage that has already been done. More broadly, advancing biorisk management relies on a continual risk assessment process to collectively build the BRM knowledge base. Simply put, encouraging and training laboratory staff to inquire and probe deeply about risks and risk assessment will ultimately strengthen the biorisk management system. A strong foundation in risk assessment will encourage the development and use of more effective mitigation and performance measures.

BRM training aims to transfer knowledge, support, skills, and abilities to critically assess risk before incidents occur. By providing opportunities to carefully evaluate the factors that contribute to risk, and effectively ask the right questions, BRM training can empower people to take responsibility for their own health, and the well-being of others and the environment. Ideally, BRM training allows for improved understanding of risk and better management of risk over time.

AN INTERNATIONAL AND HISTORICAL PERSPECTIVE

In Central Asia, the former Soviet Anti-Plague System (APS) laboratories often struggle with the best way to implement and train staff to follow the former Union of Soviet Socialist Republics (USSR) guidelines, which are still applicable to countries in the former Soviet Union (FSU). These guidelines consist of a wide set of sanitary regulations, laws, instructions, and other legal documents that ensure safety while working with especially dangerous pathogens (USSR Ministry of Health 1978, 1979), and closely resemble those adopted by Russia (Russian State Committee for Sanitation and Epidemiological Oversight 1994). They were initially established after a tragic accident where the deputy director of the Anti-Plague Institute of Saratov "Microb" became infected at work, and then traveled to Moscow for a business trip, infecting those with whom he came in contact.

The guidelines describe strict rules of working with dangerous pathogens, including detailed procedures for how to collect samples, the use of personal protective equipment, and proper disinfection, among other measures. The former Soviet republics still rely upon the information within the guidelines, although they are quickly becoming outdated with the advent of new biosafety equipment and best practices (Turegeldiyeva 2014). This poses a problem, as many of these guidelines are solidified into national law, and violating them results in strict penalties. As a result, a number of former Soviet republics are in the process of revising their legislation to include many norms and regulations from the West (Bakanidze et al. 2010; Republic of Kazakhstan 2012a, 2012b). However, this process takes time and unfortunately may not be able to provide protection to those who need it now.

The FSU countries are not alone in their drive to ensure adequate training according to current best practices and inclusion of biosafety into the legislative framework amid advances in biotechnology or challenges in infrastructure (Mtui 2011; Wang 2004). In Malaysia, biosafety education is a high priority because of the increasing prevalence of genetically modified products for which safety is a primary concern (Rusly et al. 2011). In the developing world, laboratories often struggle with a safety system dominated by engineered solutions that rely on constant electrical power, which is not always available (Heckert et al. 2011). These situation-specific challenges for bioscience facility operations translate into biorisk management training challenges. To address these challenges, laboratory managers should use a BRM training program that explicitly embraces the AMP (assessment, mitigation, performance) model and serves to specifically mitigate identified, local risks. For these reasons, BRM training must be deeply rooted in a solid foundation of risk assessment, and advancing this approach broadly will bolster positive outcomes and significantly reduce risk worldwide.

BRM training is essential to addressing increasing concerns about the effectiveness of good laboratory practices and incident reporting in laboratories given the growing number of high-containment laboratories around the world (Chamberlain et al. 2009; Ehdaivand et al. 2013). To increase BRM capabilities, various organizations have prioritized training not only to address fundamental biosafety and biosecurity issues, but also to promote an understanding of dual-use biological research (US Department of Health and Human Services 2007) and advocate for a culture of security to prevent an adverse event (Graham et al. 2008). Given this state, BRM training must create a biosafety and biosecurity culture that promotes far-reaching changes of BRM perspectives and behaviors. Therefore, we propose a BRM training paradigm shift—away from reliance on predetermined risk characterization strategies and to critical thinking training that includes a thorough risk assessment. This risk-based approach to BRM training will secure knowledge, skills, and experience in risk analysis so that individuals will make confident decisions to reduce risk, especially as biotechnology advances, new diseases emerge, and interdisciplinary work grows to meet the demands of an advancing society.

USING ADDIE FOR BRM TRAINING

This chapter will describe how recognized BRM training best practices and experiences from around the world can be used effectively to reduce risk. To demonstrate the implementation of risk-based BRM training, this chapter is structured around some core questions that are organized into a well-known instructional system design (ISD) model that incorporates five primary elements: analyze, design, develop, implement, and evaluate. This model is known as the ADDIE model (Hodell 2006) and is shown in Figure 6.1 and described in Table 6.1 in relation to BRM training. This model describes a general instructional systems design process for any training program or event. ADDIE is routinely used in various instructional settings, and has

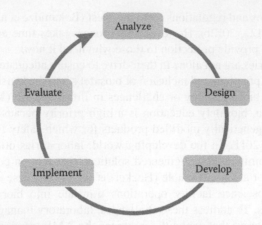

FIGURE 6.1 ADDIE cycle.

TABLE 6.1
ADDIE Cycle Components as They Relate to BRM Training

Analyze	Design	Develop	Implement	Evaluate
What is the first step to initiate BRM training?	**How can BRM training be retained?**	**What to teach?** Identifying, Compiling, and Organizing Training Content	**Who is involved in BRM training?** Understanding Student and Instructor Characteristics and Interactions	**Was the training effective?** Measuring Training Performance and Success
How are BRM training needs identified? Training Needs Assessment	Incorporating Instructional Design			
What is the next step to initiate BRM training?	**Where should BRM training take place?** Focus on the Learning Environment		**What else is needed to execute BRM training?** Other Considerations for Successful Training	
Are there general strategies for BRM training implementation? Incorporating Risk Assessment				

shown marked success in the implementation of biosafety training programs in the Ukraine, Central Asia, Kazakhstan, and Georgia (Delarosa et al. 2011).

It is important to note that the ADDIE instructional systems design model can be applied to many different instructional delivery systems. This may include face-to-face training, distance or remote training, delivered training, or any combination of face-to-face and interactive web-based training, commonly referred to as blended training. With the focus on training systems, instructional design takes into account the specific knowledge, skills, and abilities that the students have before and will acquire during the training. In other words, the ADDIE model helps ensure that the core goals and objectives of the training are met systematically regardless of the instructional delivery media.

Training Needs Assessment

The first step to initiate BRM training is to gather information about the training needs: ask questions to identify specific gaps in knowledge, understanding, skills, and abilities for a group of individuals. A training needs assessment should identify specific deficiencies or gaps in understanding or application of BRM principles that could contribute to the likelihood or consequences of an adverse event. Information may come from a number of sources, including those listed in Table 6.2. Many of the same tools used for risk assessment will also identify training needs if the information is tailored toward a training perspective. One particularly useful method is to analyze job descriptions or responsibilities within an institute. For BRM to be most effective, everyone must perform his or her duties effectively. Therefore, a workforce analysis can help identify those individuals who have specific roles, or specific knowledge, skills, and abilities necessary to achieve BRM goals and objectives. This is a fairly simple way to determine where training could be directed.

For example, using international guidance as a starting point, Delarosa et al. (2011) divided up all the relevant skills necessary to perform a biosafety-related job into tasks and subtasks. Then, they assessed each position's biosafety proficiency using performance metrics (Figure 6.2). This information became the basis for implementing biosafety training programs. Similarly, the guidelines for biosafety competency, drafted by the CDC, can be used to align BRM training efforts to objectives that are common to many organizations' safety programs (US Centers for Disease Control and Prevention 2011). The competencies are divided into four domains with specific topics and skill levels (entry, mid, senior). These guidelines are an excellent resource not only to assess current training, but also to help direct

TABLE 6.2
**Information Sources for a Training
Needs Assessment**

Risk assessment

Job description analysis

Root cause analysis

Surveys

Interviews

Historical data

Behavioral observation data

Laboratory inspection

Drills

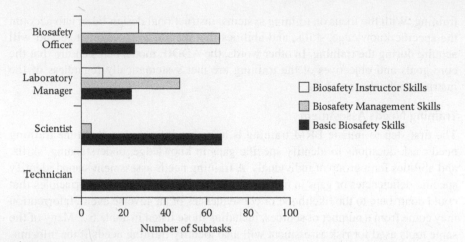

FIGURE 6.2 Needs assessment tools—focus on job description analysis. (Data from Delarosa et al., *Applied Biosafety: Journal of the American Biological Safety Association*, 16, 2011.)

and develop training programs. This approach for assessing training needs should be directly linked to the risk assessment. Surveys are another powerful training needs assessment tool that can be easily customized for a variety of training situations. Some examples of how surveys have been used to gather information to better understand training needs are included in Table 6.3 (Ehdaivand et al. 2013; Kahn 2012; Nasim et al. 2012).

Whatever method is used, the information obtained from a training needs assessment should be cross-referenced with an institution's overall risk assessment. Integrating performance indicators and regular testing into the training needs assessment ensures that the results can be easily measured over time and documented to demonstrate risk reduction. These examples highlight the advantage of directed BRM training, which targets the root cause of the risk compared to general, nonspecific biosafety training.

Other factors, such as national and international guidance, or regulatory or legal requirements, can also shape BRM training. High-level policy decisions will often shape BRM training requirements for an institute, laboratory, or in some cases, a country. International BRM guidance from the CWA 15793 and CWA 16393 documents outlines general components of a biorisk management system and implementation guidelines, respectively (European Committee for Standardization 2011, 2012). See Table 6.4. The CWA 15793 has also been used successfully as a gap analysis tool to identify BRM needs, including training (Rengarajan 2010; Sundqvist et al. 2013). Additional guidance documents from the World Health Organization (WHO), the US National Institutes of Health (NIH), the US Centers for Disease Control and Prevention (CDC), and other institutions are included in Table 6.5. A thorough review of the literature is often a good starting point to shape BRM training policy.

The consequences of not having a BRM training program may be noncompliance depending on the guidelines and legal requirements of a laboratory, institute,

TABLE 6.3
Needs Assessment Tools—Focus on Surveys

Description	Outcome	Country	Reference
Survey of laboratory practices along with a review of current laboratory guidelines to understand if biosafety practices in anatomical pathology laboratories are sufficient.	Favorable perceptions of adequate biosafety training did not reflect daily practices or current guidelines.	United States	Ehdaivand et al. 2013
Survey of graduate students and postdoctoral fellows from large public and private universities to assess students' education, experiences, and attitudes regarding biosecurity in comparison to bioethics or biosafety.	The concept of biosecurity was new to many respondents compared to bioethics and biosafety. There is a lack of biosecurity training opportunities.	United States	Kahn 2012
Survey of laboratory technicians from clinical laboratories in public and private sector hospitals to determine biosafety perception and practices.	The majority of laboratory technicians had not had any formal biosafety training and lacked awareness of good laboratory practices and biosafety measures amid poor resources.	Pakistan	Nasim et al. 2012

locality, or country. Knowledge and understanding of specific legislation are essential for any BRM training program.

Currently, there are no internationally recognized BRM training standards or certified BRM professionals, although national biosafety associations, such as ABSA, offer biosafety credentials. In addition, the International Federation of Biosafety Associations (IFBA) is in the process of creating an internationally recognized BRM professional certification (International Federation of Biosafety Associations 2012). Commonly, each institute, in conjunction with its institutional biosafety committee and biosafety professional, will work to establish its own training standards. Exactly how these policies, guidelines, recommendations, and standards are incorporated into BRM training is often the role of the institute's management and leadership in association with a biosafety advisor or other BRM training professionals.

The training needs assessment should be continually updated. Revision should occur whenever there is new information, such as results from a previous BRM training event, or if the conditions or situation in the facility changes. For example, if a new laboratory, new person, new piece of equipment, or even a new responsibility is added to the work environment, the training needs should be reevaluated. Overall, the process of reducing risk through training is a moving target that needs continual assessment to determine how biosafety and biosecurity training can most effectively be used to reduce risk.

TABLE 6.4
BRM Training Guidance

CWA 15793:2011
4.4.2.4 Training

"The organization shall ensure that requirements and procedures for biorisk-related training of personnel are identified, established, and maintained.

NOTE Procedures should address:

a) definition of biorisk training needs;

b) provision of required biorisk training;

c) determination of effectiveness of biorisk training;

d) provision of refresher biorisk training;

e) restrictions on personnel to ensure they do not perform tasks for which they are not trained;

f) maintenance of adequate records.

Training should include raising personnel awareness of biorisk issues including the relevance of human factors in biorisk management."

CWA 16393:2012
4.4.2.4 Training

"The organization should design, establish, maintain, assess and monitor a robust biorisk-related training programme appropriate to all levels of personnel. The training should include raising personnel awareness of biorisk issues including the relevance of human factors (e.g. behaviour, reliability, errors) in biorisk management.

The result should be a training programme that provides staff with the knowledge and skills to reduce risk, with measurable markers of success that can be reported to management.

In order to design a training programme the organization should consider—apart from the ones already given:

- definition of employees duties and their training needs related to biorisk (e.g. analysis of employees past performance);
- biorisk policies and objectives;
- safety and security competencies that are required at different levels of the organization;
- provision of required biorisk training;
- frequency of training;
- awareness programmes for contractors, temporary workers and visitors;
- determination of effectiveness of biorisk training;
- choice of the appropriate method for conducting the training (e.g. web based, instructor led, hands-on training);
- restrictions on personnel to ensure they do not perform tasks for which they are not trained; and
- documentation and maintenance of adequate records of training that include attendance and content of training.

Training programmes should equip staff with the knowledge and tools to identify hazards, manage risk, and put into place measurable markers of success that can be reported to and used by management."

Source: European Committee for Standardization, "CEN Workshop Agreement (CWA) 15793: Laboratory Biorisk Management," 2011; European Committee for Standardization, "CEN Workshop Agreement 16393:2012: Laboratory Biorisk Management—Guidelines for the Implementation of CWA 15793:2008," 2012.

TABLE 6.5
Sources for BRM Training Guidance

Source	Description	Website
CDC	Centers for Disease Control and Prevention	http://www.cdc.gov/biosafety/
		http://www.cdc.gov/mmwr/pdf/other/su6002.pdf
	Guidelines for laboratory competency	http://www.selectagents.gov/resources/Guidance_ for_Training_Requirements_v3-English.pdf
	Select agent guidance	
CWA 15793	CEN Workshop Agreement, 2011	ftp://ftp.cenorm.be/CEN/Sectors/TCandWorkshops/ Workshops/CWA15793_September2011.pdf
CWA 16393	CEN Workshop Agreement, 2012	ftp://ftp.cen.eu/CEN/Sectors/List/ICT/Workshops/ CWA%2016393.pdf
CDC/NIH	Biosafety in microbiological and biomedical laboratories	http://www.cdc.gov/biosafety/publications/bmbl5/ bmbl.pdf
NSABB	National Science Advisory Board for Biosecurity	http://osp.od.nih.gov/office-biotechnology-activities/ biosecurity/nsabb
NIH	National Institutes of Health	http://osp.od.nih.gov/office-biotechnology-activities/ biosafety
NBBTP	National Biosafety and Biocontainment Program	http://www.nbbtp.org
OSHA	Occupational Health and Safety Administration	https://www.osha.gov/SLTC/laboratories/index.html
CLSI	Clinical Laboratory Standards Institute	http://clsi.org
WHO	World Health Organization— *Biosafety Manual: Laboratory Biosecurity Guidance*	http://www.who.int/csr/resources/publications/ biosafety/WHO_CDS_CSR_LYO_2004_11/en/
		http://www.who.int/csr/resources/publications/ biosafety/WHO_CDS_EPR_2006_6.pdf
SNL	*Laboratory Biosecurity Handbook*	Salerno and Gaudioso 2007
IHR	International Health Regulations, 2005	http://whqlibdoc.who.int/publications/2008/ 9789241580410_eng.pdf
ASM	American Society for Microbiology—*Guidelines for Biosafety in Teaching Laboratories*	http://www.asm.org/images/asm_biosafety_ guidelines-FINAL.pdf
		http://www.asm.org/images/Education/FINAL_ Biosafety_Guidelines_Appendix_Only.pdf

Incorporating Risk Assessment

Using BRM training as a risk mitigation measure assumes an understanding of the risk and how training will address that risk. In other words, the results of the risk assessment should be used to draft specific BRM training goals and objectives, and the focus should be on reducing unacceptable risks. Goals should be broad in scope and incorporate the various elements of the AMP model. BRM objectives should define more specific, action-oriented outcomes for a particular BRM training. Depending on the organization, the terminology for goals and objectives may vary. Table 6.6 includes an example of training goals and objectives derived from training

TABLE 6.6
Example of a Learning Outcomes Approach in Egypt

General Goals Addressed	Specific Learning Objectives/Outcomes	Types of Assessments That Measure Objective	Activity That Accomplishes a Specific Objective
Participants will be advocates for teaching responsible conduct of research and practice of science.	Develop a teaching module to illustrate the use of the concepts of responsible conduct of research.	Develop an assessment instrument that will demonstrate the student's ability to use the concepts to solve practical problems. Use a historical case study to engage students and deepen awareness of the various issues.	Present your approach to your colleagues in the institute and obtain their feedback.
Participants will have an awareness of hazards in the laboratory and know how to bring that awareness to others.	Identify the difference between chemical and biological hazards. Be able to describe biosafety guidelines and standards of practice to prospective trainees.	Tested knowledge, pre- and post-assessment. Offer a problem and ask students to describe any obvious hazardous situations.	Group activities, small group discussions, clicker questions. Expertise sharing (own experiences of best practice, own stories of poor practices).
Participants will appreciate the ethical, legal, and social responsibilities of life scientists.	Identify policies and guidelines and regulatory statements of both international and local bodies and critique the applicability of these statements. Be able to write standards of practice for their own institution, department, or laboratory.	Convey these policies to their peers/ colleagues in their native language. Critique and discuss how these apply to participants' own experience, laboratory, institution, or country.	Locate and read/ discuss these guidelines with the group. Discuss cases from historical examples (e.g., Thomas Butler). Discuss case studies specific to the group itself (e.g., based on personal experience).

Source: National Research Council, *Research in the Life Sciences with Dual Use Potential: An International Faculty Development Project on Education about the Responsible Conduct of Science*, National Academies Press, Washington, DC, 2011.

initiatives that focused on the responsible conduct of research in Egypt (National Research Council 2011). Training development should always take into account the various institutional, student, and instructional training goals and objectives. It is a natural tendency for instructors to want to deliver as much information as possible. However, if some information does not directly correlate to the goals and objectives, including it may dilute the specific risk mitigation benefits of the training.

Understanding the scalable nature of BRM training goals and objectives allows the training to be implemented for a number of different roles and organizational levels in a way that can have far-reaching effects on risk reduction. The goals and objectives for BRM training must acknowledge preexisting knowledge, skills, and abilities of the students and their functional roles. All personnel roles, whether they are top management, policy makers, representatives of funding organizations, students, laboratory workers, security staff, or the cleaning crew, can influence and impact biorisk management. BRM training goals and objectives should focus on augmenting the specific knowledge, skills, and abilities of specific personnel to enhance their ability to mitigate the risks identified in the risk assessment.

There are a number of training strategies that can be used to achieve BRM training goals and objectives. Many of these strategies will depend on the institute and the context of the training. For example, biosafety training programs in high-containment laboratories are often structured to transfer knowledge through a mentor-apprentice relationship. Through one-on-one training and experience, the student gradually achieves greater independence (Gronvall et al. 2007). Mentorship, as with other training strategies listed here, has inherent advantages and disadvantages (Lee et al. 2007; Walker et al. 2002). The training strategy may be topic specific, role specific, or country specific. There are a number of examples from the literature where various training strategies have been implemented, such as developing responsible science teaching capacities in the Middle East and North Africa (National Research Council 2013), strengthening research capacity for infectious diseases in Honduras (Sanchez et al. 2012), or preparing hospital staff to treat potential exposures from a high-containment lab in the United States (Risi et al. 2010).

The literature describes the successes and failures of many previous training initiatives. Each strategy has advantages and disadvantages that should be considered prior to initiating any training, which is a key component of the analyze step in ADDIE. It is also important to be aware that there may be cultural considerations that would favor one training strategy or approach over another.

INCORPORATING INSTRUCTIONAL DESIGN

It is well understood that learning does not happen instantly; instead, learning must be built upon a foundation of knowledge and experience. This progression is best described in Bloom's taxonomy (Anderson and Krathwohl 2001; Bloom 1956), which is depicted in Figure 6.3. In this model, learning occurs in five stages, starting from a basic foundation of information and progressing to more advanced thought processes: know, learn (comprehend), apply, analyze, evaluate. Bloom's taxonomy is useful to clearly define BRM training goals and objectives and to design training accordingly. This model has been used frequently in structuring science and

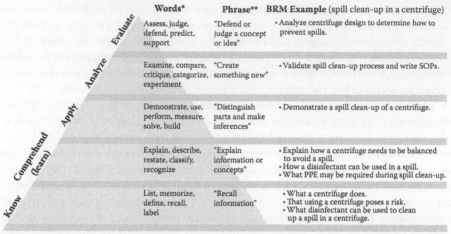

	Words*	Phrase**	BRM Example (spill clean-up in a centrifuge)
Evaluate	Assess, judge, defend, predict, support	"Defend or judge a concept or idea"	• Analyze centrifuge design to determine how to prevent spills.
Analyze	Examine, compare, critique, categorize, experiment	"Create something new"	• Validate spill clean-up process and write SOPs.
Apply	Demonstrate, use, perform, measure, solve, build	"Distinguish parts and make inferences"	• Demonstrate a spill clean-up of a centrifuge.
Comprehend (learn)	Explain, describe, restate, classify, recognize	"Explain information or concepts"	• Explain how a centrifuge needs to be balanced to avoid a spill. • How a disinfectant can be used in a spill. • What PPE may be required during spill clean-up.
Know	List, memorize, define, recall, label	"Recall information"	• What a centrifuge does. • That using a centrifuge poses a risk. • What disinfectant can be used to clean up a spill in a centrifuge.

*Bloom 1949
**From NRC "Developing Capacities for Teaching Responsible Science in the MENA Region" 2013

FIGURE 6.3 Active learning models and strategies: Bloom's taxonomy (from Bloom, *Taxonomy of Educational Objectives: The Classification of Educational Goals*, 1st ed., Longmans, Green, New York, 1956; National Research Council, *Developing Capacities for Teaching Responsible Science in the MENA Region: Refashioning Scientific Dialogue*, National Academies Press, Washington, DC, 2013).

medical professional training (National Research Council 2011, 2013; World Health Organization 2006). Most introductory and BRM awareness training falls into the *know* to *apply* stages. Higher-order cognition, in the *analyze* and *evaluate* stages, often targets biorisk management professionals, who have responsibility for implementing a complete biorisk management system. Generally, it is not reasonable to expect a student to progress more than one taxonomy level in any one training session. In some cases, it may take multiple training sessions or courses for students to advance to the next level.

The Centers for Disease Control and Prevention uses an outcome-based training model that relies upon specific roles and competencies to drive public health training initiatives (Koo and Miner 2010). This so-called Dryfus model complements BRM training because many of the training objectives of public health institutes overlap with core biosafety and biosecurity principles. The focus in designing BRM training should be on framing the content in a way that will result in an effective learning experience.

Advancements in human cognition and learning research have resulted in a deeper understanding of how people learn. Furthermore, the scientific disciplines rely upon a deep understanding of a topic, rather than just basic knowledge. Therefore, over the last quarter century, science education has pushed for teaching pedagogy that capitalizes upon how people learn naturally (e.g., learning to walk, speak, ride a bicycle, or play a musical instrument) to achieve deep understanding of scientific topics (National Research Council 2000; Tanner and Allen 2004). Variations of individual student learning styles explain why some learners understand best if information is shown, demonstrated, or performed, while others may understand best if key information is heard or read (Freeman et al. 2014). Recent research challenges traditional lecture-style learning environments to become more student centered and interactive in nature. Interactive learning includes visual or physical demonstrations, videos, case studies, role-playing, group activities, problem solving, and games. Active learning has been proven to increase student performance in science, engineering, and mathematics (Freeman et al. 2014).

Traditional learning theory promotes a content- and instructor-driven approach in which information is primarily explained or shown to students by the instructor. In this situation, students have the responsibility to retain the information provided by the instructor. However, in association with advances in understanding human cognition, learning theory has evolved to a more facilitated- and learner-based approach where content is paired with carefully designed learning aids, activities, environments, and situations. In this approach, it is the responsibility of the instructor to guide student inquiry along a path of information self-discovery (National Research Council 2000). Generally, learner-based approaches, also known as facilitated learning, utilize more of a dynamic, interactive type of knowledge transfer.

A primary driver for the shift to a learner-based approach is an advanced understanding of how learning and memory are interconnected, as well as the observation that students exposed to active learning outperform students taught with more traditional methods (Freeman et al. 2014). Furthermore, undergraduate university professors and administrators are taking note that active learning correlates to not only increased student performance, but also increased instructor interest and satisfaction (Armbruster et al. 2009).

There are a number of learning models that support the learner-based approach, including the principles of learning (Figure 6.4) and Jensen's model (Figure 6.5). The principles of learning lists strategies to design and deliver training in a memorable

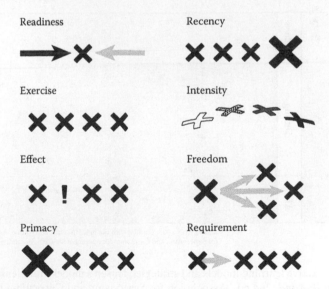

Readiness

Recency

Exercise

Intensity

Effect

Freedom

Primacy

Requirement

FIGURE 6.4 Active learning models and strategies: Principles of learning (from Thorndike, *The Fundamentals of Learning*, Teachers College Press, New York, 1932).

way, encouraging instructors to incorporate proven learning design features into curriculum development (Thorndike 1932). Key definitions for the principles of learning are:

1. Readiness: Students learn best when they are mentally, physically, and emotionally ready to learn, including having a strong motivation to learn.
2. Exercise: Students will remember key points if they are repeated.
3. Effect: Learning is strengthened by a positive and impactful learning experience.
4. Primacy: The material learned first is often remembered.
5. Recency: The material most recently learned is often remembered.
6. Intensity: Intense and stimulating material, which may include activities, demonstrations, etc., is more likely to be retained.
7. Freedom: The student will invest effort in learning items that he or she chooses.
8. Requirement: Students learn best when starting from a common point.

The Jensen model categorizes the training effort so that instructors can focus attention, resources, and design to favorably impact learning, memory, and retention (Jensen 2005, 2008; Hodell 2006). This model relies heavily on the dynamic interactions between the student and the instructor during instruction to frame and elaborate upon the key messages. During this process, the students acquire knowledge through engaging learning experiences that allow them to connect the messages to other information, so that the content is retained. Taken together, both the principles of learning and Jensen's model give instructors resources to build upon tendencies in human cognition to make the most of instructional time and effort.

10%	Before	PREPARE your learners CREATE an optimal environment
80%	During	ENGAGE learners by getting them vested emotionally FRAME learning to make it relevant, important, and compelling ACQUIRE knowledge, skills, values, experiences ELABORATE and deepen the learning through trial-and-error time with feedback and active processing CONNECT learning to other content, processes, and self
10%	After	SETTLE the learning with time for passive processing REHEARSE & INCORPORATE by revising learning and using it

Jensen, Eric. 2005. Teaching with the brain in mind, 2nd edition. Association for Supervision and Curriculum Development (ASCD): Alexandria. Virginia

FIGURE 6.5   Active learning models and strategies: Jensen's model (from Jensen, *Teaching with the Brain in Mind*, 2nd ed., Association for Supervision and Curriculum Development (ASCD), Alexandria, VA, 2005).

The traditional learning methodology based on strict lecture, while definitely informative, often requires students to integrate knowledge and skills outside of the classroom—an expectation that is often not realistic for BRM training given time, resources, and other constraints. A perceived drawback to facilitated learning is that more time is spent on teaching only a few key messages, whereas in traditional instructor-based approaches, many more key messages can be conveyed in the same amount of time. However, evidence shows that when memory and retention are challenged, overall learning in active learning environments is greater than in instructor-based settings (Freeman et al. 2014). These learning models help instructors to understand that smaller amounts of meaningful and relevant information will achieve more learning and retention than large quantities of information on any particular subject.

A distinct advantage of facilitated learning, especially for BRM training, is that it provides information about the students to the instructor. In this way, the instructor may identify additional BRM training needs or other gaps that need to be filled or incorporated into the risk assessment. With one-way communication only from teacher to student, this is not possible. Furthermore, facilitated learning promotes camaraderie between the students and the instructor because everyone works together to achieve common goals, such as improving BRM. As a result, the group can collectively bring different perspectives, experiences, and abilities to the learning environment to help improve biosafety and biosecurity.

Focus on the Learning Environment

BRM training can take place anywhere, although there are settings and considerations for the learning environment that can impact the training event and, consequently, student learning. For example, Maslow's hierarchy of needs (Maslow 1943) states that basic physiological needs (oxygen, water, food, sleep) and safety (security, job, family, health, property) must be met before learning can take place. Taking this information into account can help an instructor maximize the learning environment, including the student's motivation to learn (Raffini 1993). For example, if students are extremely hungry, tired, or even uncomfortable, they will not be able to absorb the key messages of the training. Therefore, allowing time for coffee breaks, snacks, and even a chance for students to quickly check their email will increase learning, and also help the students to enjoy the training more.

Similarly, an awareness of the student's culture and the region of the world where the training is being conducted will help the instructor plan the training according to local customs and expectations (National Research Council 2000). In addition, effective BRM training relies heavily on communication to convey key messages; therefore, extra effort must be made to ensure that proper translation and interpretation are available when warranted. Also, it is critical to understand that the words for *biosafety* and *biosecurity* may be similar or even the same word in some languages, or the concepts may be completely new. As a result, extra time may be needed to clarify key terms and definitions before more advanced topics are covered. Ideally, international BRM training fosters cross-cultural exchange of BRM best practices that benefit the BRM community as a whole.

Besides creating an optimal training environment and accounting for cultural and language issues, BRM training must consider the venue. Arguably, the most effective BRM training is done in the environment where the hazard is located. This is often a laboratory, but it may be in a field environment, a hospital, or even an infectious substances transport system. For example, the laboratory network Biotox-Piratox used a series of exercises to assess the bio-preparedness and the training of French hospital laboratories in the event of a biological threat. Their methodology focused around four exercises between 2007 and 2011, where correct diagnostic and reporting procedures were tested (Merens et al. 2012). Even though these exercises were not specifically for training purposes, drills and other real-world simulations can have a profound impact on learning. This has been coined experiential learning (Kolb et al. 1984). Kaufman and Berkelman (2007) show an example of experiential biosafety learning in the high-containment environment, using a mock laboratory.

Such hands-on training gives students the ability to actively assess and mitigate risks, and apply performance evaluations. Unfortunately, resources, time, and other constraints often limit the extent to which this learning environment can be utilized. Informal learning contexts can be great alternatives. For example, a panelist session can be arranged during lunch where experts can share their knowledge and expertise on a particular subject. BRM training in other learning environments, such as in social media (Facebook, Twitter, Weibo, or LinkedIn), can also be effective if it incorporates design and development elements that promote memory and learning.

IDENTIFYING, COMPILING, AND ORGANIZING TRAINING CONTENT

The next step is to consider what BRM content would be best suited to mitigate the identified risk. At this stage, the instructor should draft key messages that align with the overall goals and objectives of the training. The process of drafting key messages requires reviewing the relevant information and literature, and then correlating that information to the training objectives that were initially established. Other considerations regarding BRM training design should be accounted for so that time is focused on developing activities and learning experiences that target the most relevant key messages. The act of drafting key messages will direct the acquisition, processing, and development of the training materials and content in a way that not only facilitates instructor knowledge and breadth, but also provides a solid foundation for training design, delivery, and development.

Key messages are often specific to a particular BRM topic. They may fall in categories such as infectious substance shipping and transport, incident response, project management, equipment and engineering controls, laboratory design, laboratory operations, and maintenance, to name just a few. There are a number of excellent resources for BRM content. Some of these are listed in Table 6.7. These resources can be used to build a reference library from which BRM training content can be developed over time to address changing BRM needs. There are a number of efforts within the biosafety community to collect training material in a central location. However, it is important to note that no single source of BRM content will fit all BRM needs.

TABLE 6.7

Additional Sources for Biorisk Management Content

Description	Website
CDC laboratory training resources	www.cdc.gov/labtraining/index.html
American Biological Safety Association training tools	http://www.absa.org/trainingtools.html
University of Texas Medical Branch National Biocontainment Training Center	www.utmb.edu/nbtc/
Sandia National Laboratories, Global Biorisk Management Curriculum	http://www.biosecurity.sandia.gov/gbrmc/index.html
American Society for Microbiology	http://www.asm.org/index.php/educators/curriculum-guidelines

As much as possible, it is good to integrate locally relevant and situation-specific information when developing the content for a training event. Often local information is not readily available in written form. It may come from direct or indirect experience of working in a local laboratory. Utilizing local risk assessments as the basis for BRM training is often the best method for developing training that reflects the unique aspects of a particular laboratory.

Perhaps the greatest challenge in creating BRM content is balancing between high and low containment settings, high and low resource environments, fieldwork and laboratory work, and even biosafety and biosecurity. Although this book references international best practices frequently, published best practices cannot account for every conceivable situation. The AMP model, however, applies to all biosafety and biosecurity scenarios, and should always be the basis of training design.

Once the BRM training content is finalized, the content needs to be organized in a way that is meaningful for the student. It should have a consistent flow and a good mix of activities and opportunities for the students to learn effectively. Rather than only focusing on theory, the content should include class time for work to develop specific deliverables. The type of deliverable should correspond with the desired Bloom's taxonomy learning level for the training (Bloom 1956). The quality of these deliverables may also serve as an element of evaluation to determine the extent of learning.

Organizing the flow and delivery of the content is the last step to finalize the materials for the course. This will blend together elements from both the design and develop stages of ADDIE to form a clear lesson plan and instructor guide, which includes relevant background information, reference material, and expected responses from facilitated group activities. Documenting this information will help novice instructors to prepare to teach the material. Also, organizing the course material will structure the training in a way that keeps the content on track. Finally, this documentation will serve as a foundation for continual improvement of the course materials. After developing a training course for the first time, the instructors should conduct pilot testing of the course to make sure that all elements of the training work together to meet the desired objectives.

Understanding Student and Instructor Characteristics and Interactions

Students

Although many people inside and outside of an institute can benefit from BRM training, those who work with potentially dangerous pathogens should be the primary targets of BRM training. Typical roles include laboratory workers, lab managers, principal investigators, students, and even law enforcement officials, emergency responders, and visitors to the laboratory. Executive management at an institute should also receive training to help direct biorisk management policy decisions. Each person may have a unique background and a different level of risk aversion or tolerance, which may influence his or her level of interest in the training.

Especially when teaching adult learners, the instructors should ask students about their experience, role, interests, and current knowledge of BRM. This engagement will help to align the targeted audience with the most beneficial coursework (National Research Council 2000). Ideally, this information should be gathered before the training begins in a training needs assessment. It can also be gathered at the beginning of a training using a pretest or by interviewing the students alone or in groups. What do they do? Why are they taking the training? What do they want to learn? This strategy helps to manage expectations between the students and the instructors before the class begins, which allows both parties to adjust accordingly. This is especially important in settings where professional networking or self-selection drives participation. In general, it is much more effective to identify the gaps in the BRM system, and then select students based on their roles and responsibilities.

Instructors

BRM is most effectively taught with an experienced instructor who can manage all the aspects of the ADDIE cycle and also has a comprehensive understanding of biorisk management. Other important instructor characteristics include the following: respectful, flexible, culturally aware, able to convey information and answer questions clearly, interesting, fun, and sympathetic (Marsh 1980). A well-rounded instructor can take years to develop. Large institutional BRM programs should invest significant resources in developing effective BRM trainers. BRM training performance metrics can help track instructor qualifications and experience, and student evaluation feedback can help determine an instructor's strengths and weaknesses. Less experienced instructors can be teamed with more experienced instructors as often as possible to constantly grow the cadre of an institution's trainers.

In general, the responsibility for conducting BRM training falls upon the shoulders of an institutional biosafety professional or a laboratory manager. Regardless of who conducts the training, the instructor must have a comprehensive understanding of the risk assessment results. At a minimum, the BRM trainer needs to be knowledgeable in risk assessment and the facility operations.

The instructor needs experience in classroom management and facilitating learning, especially in the dynamic environment of experiential learning activities. In learner-centered approaches, the students take control of their own learning and they will need time to work through activities, questions, or problems, with the instructor guiding them along the way. This is especially effective in BRM training, as many skills require critical thinking abilities to evaluate different scenarios in the classroom. Unfortunately, there are not many opportunities to develop learner-centered types of facilitation skills besides practice and co-teaching with a more experienced instructor. A few funding organizations, such as the World Health Organization (WHO), the US Department of State's Biosecurity Engagement Program (BEP), the US Cooperative Biological Engagement Program (CBEP), the American Biological Safety Association (ABSA), and the International Federation of Biosafety Associations (IFBA), have promoted this type of BRM learning internationally, and offer learning opportunities specifically for BRM instruction.

At present, there are no internationally recognized BRM instructor credentials or certifications. Instead, management and leadership at individual institutes or organizations determine BRM trainer qualifications. The CDC biosafety core competencies identify knowledge that an instructor should have about biosafety topics, but these competencies may not be universal (US Centers for Disease Control and Prevention 2011). This is important to understand when developing institutional training programs or obtaining BRM training from an outside organization.

Other Considerations for Successful Training

A common statement is: "If we had more time and resources, we would increase the amount of training we would provide to our employees." Therefore, focusing BRM training to address priority items identified within the risk assessment is the best use of limited resources. Time and budget should be seriously considered in the initial planning stages of BRM training to scope the goals and objectives of the training program or training event. Careful planning to match the training audience with the appropriate subject matter, combined with a delivery designed for high student retention, can address most time and budget constraints.

MEASURING TRAINING PERFORMANCE AND SUCCESS

Evaluation is critical for determining if the training was effective or not. There are many levels, options, and strategies for measuring BRM training success. Creating student and course evaluations that focus on the key messages is an effective way to determine if the learning objectives were achieved. For example, BRM training may incorporate a pre- and post-test, just a post-test, a survey, a homework assignment, rubrics, or even behavioral observation data as a way to determine if the training was effective.

Overall, there is not a standard approach to evaluate training, yet it is important to recognize that this evaluation data can have a significant impact on BRM as a whole for an institute. Training programs in other fields rely extensively upon four levels of evaluation proposed by Kirkpatrick (2007) and depicted in Table 6.8. Hodell (2006) addresses how these levels are easily integrated into instructional systems design for a training program. Generally, the first level measures student reaction to a training course in terms of likes and dislikes (e.g., Did the course meet my expectations? Was the training environment adequate? Was the instructor knowledgeable?). The second level focuses on learning, more specifically: Were the learning objectives for the course met? At this level, the objectives themselves may serve as the media for

TABLE 6.8
BRM Training Evaluation

Level 1   Was the student *happy* with the course?
Level 2   Did the student *learn*?
Level 3   Over time, did the student's *behavior* change to meet the desired objective?
Level 4   Over time, did the *organization see improvement* in biorisk management?

Source: Kirkpatrick, *The Four Levels of Evaluation: Measurement and Evaluation,* Alexandria, VA: American Society for Training & Development, 2007.

evaluation. For example, for a course on personal protective equipment (PPE) use, an objective may be "Students will be able to select appropriate PPE based on route of exposure of a microorganism." Completion of this objective may be asked from either or both the student's and instructor's perspective using a Likert scale, or other means, and then analyzed. The third level measures behavior. This typically includes how much of the training is transferred to the workplace and can be measured using surveys, observations, interviews, or other means designed specifically to the context and application of the training. For example, for PPE training this may include a laboratory site visit where behavioral observation data are used to determine the number of staff who wear the appropriate PPE. The fourth level is results, or organizational change at the institute level. At this level, the bottom-line results are measured. For BRM, this is risk reduction, which, arguably, is hard to measure. When used in combination with institute performance measures, and risk assessment data over time, this risk reduction question can be answered.

Unfortunately, in many institutions, evaluation and performance metrics do not exist or are incomplete. Monitoring BRM training performance is critical for determining what topics need further training and what teaching methods were most effective. Having a complete record that tracks who received training, what they were taught (according to the objectives), and if the objectives were achieved will help determine how much training contributes to reducing an institute's risks. It is also useful to collect evaluation data from a number of different perspectives. Input and feedback about BRM training from the instructor, students, management, funding partner, observers, and independent stakeholders should all be considered.

In the event that the post-event analysis shows that the training objectives were not met, then it is necessary to determine exactly why. Sometimes it may be due to the timing of the course, the instructor, the environment, the course material, the activities, or even the audience. Identifying the precise reasons why the objectives were not met will help focus and improve future training. Moreover, documenting training results over time allows for trend analysis, which will improve future course design and implementation. Therefore, understanding how BRM training can affect BRM culture can dramatically reduce the number of laboratory-acquired infections, losses, thefts, and diversions from the laboratory. The continuous analysis of BRM training effectiveness should also be a factor specifically considered by institution risk assessments. In this way, BRM training can have a direct impact on risk reduction.

CONCLUSION

BRM training is often considered one of the more sustainable risk mitigation measures because knowledge, skills, and abilities can be passed from one person to another. In addition, BRM training is cost-effective compared to other risk mitigation measures, such as large-scale engineering controls or facility equipment. BRM training has recently been the beneficiary of a growing network of formal and informal resources that have been used to better understand biorisk and direct BRM training efforts. Within the last 10 years, BRM has become a more prominent topic in undergraduate and graduate curricula around the world, suggesting that the BRM knowledge base will continue to grow. In contrast to other scientific disciplines, BRM is not an established research-based or degree-based field. Fortunately, there are a few international, regional, national, and local organizations, regulatory bodies, and experts with a vested interest in promoting BRM and providing resources to supplement efforts for BRM training. As the community of skilled BRM professionals grows, the number of people who will have an opportunity to take advantage of effective BRM training will also grow.

The future BRM landscape must inevitably tackle a number of safety concerns associated with advances in biotechnology and interdisciplinary scientific work. While these fields have the promise of advancing society in ways never imagined, there is always the possibility that this same technology could be misused accidentally or intentionally. A continuous dialogue on biosafety and biosecurity as new technology is developed will minimize this possibility. BRM training and awareness raising can, and should, play a lead role in influencing future decisions regarding ethics, safety, and security of a rapidly evolving scientific field.

REFERENCES

Anderson, L.W., and D.R. Krathwohl. 2001. *A Taxonomy for Learning, Teaching, and Assessing: A Revision of Bloom's Taxonomy of Educational Objectives.* Complete ed. New York: Longman.

Armbruster, P., M. Patel, E. Johnson, and M. Weiss. 2009. Active Learning and Student-Centered Pedagogy Improve Student Attitudes and Performance in Introductory Biology. *CBE Life Sciences Education*, 8: 203–213.

Bakanidze, L., S. Tsanava, and N. Tsertsvadze. 2010. Biosafety and Biosecurity in Georgia: New Challenges. *Applied Biosafety: Journal of the American Biological Safety Association*, 15: 85–88.

Bloom, B.S. 1956. *Taxonomy of Educational Objectives: The Classification of Educational Goals.* 1st ed. New York: Longmans, Green.

Chamberlain, A.T., L.C. Burnett, J.P. King, E.S. Whitney, S.G. Kaufman, and R.L. Berkelman. 2009. Biosafety Training and Incident-Reporting Practices in the United States: A 2008 Survey of Biosafety Professionals. *Applied Biosafety: Journal of the American Biological Safety Association*, 14: 135–143.

Delarosa, P., S. Kennedy, M. Jones, M. Kinsey, and L. Gardiner. 2011. Quantifying Competency in Biosafety: Adaptation of the Instructional Systems Design Methodology (ISD) to Biosafety and Laboratory Biosecurity. *Applied Biosafety: Journal of the American Biological Safety Association*, 16.

Ehdaivand, S., K.C. Chapin, S. Andrea, and D.R. Gnepp. 2013. Are Biosafety Practices in Anatomical Laboratories Sufficient? A Survey of Practices and Review of Current Guidelines. *Human Pathology*, 44: 951–958.

European Committee for Standardization. 2011. *CEN Workshop Agreement (CWA) 15793: Laboratory Biorisk Management.*

European Committee for Standardization. 2012. *CEN Workshop Agreement 16393:2012: Laboratory Biorisk Management—Guidelines for the Implementation of CWA 15793:2008.*

Freeman, S., S.L. Eddy, M. McDonough, M.K. Smith, N. Okoroafor, H. Jordt, and M.P. Wenderoth. 2014. Active Learning Increases Student Performance in Science, Engineering, and Mathematics. *Proceedings of the National Academy of Sciences of the United States of America*, 111: 8410–8415.

Graham, B., J. Talent, and G.T. Allison. 2008. *World at Risk: The Report of the Commission on the Prevention of WMD Proliferation and Terrorism.* 1st Vintage Books ed. New York: Commission on the Prevention of Weapons of Mass Destruction Proliferation and Terrorism (U.S.), Vintage Books.

Gronvall, G.K., J. Fitzgerald, A. Chamberlain, T.V. Inglesby, and T. O'Toole. 2007. High-Containment Biodefense Research Laboratories: Meeting Report and Center Recommendations. *Biosecurity and Bioterrorism: Biodefense Strategy, Practice, and Science*, 5: 75–85.

Heckert, R.A., J.C. Reed, F.K. Gmuender, M. Ellis, and W. Tonui. 2011. International Biosafety and Biosecurity Challenges: Suggestions for Developing Sustainable Capacity in Low-Resource Countries. *Applied Biosafety: Journal of the American Biological Safety Association*, 16: 223–230.

Hodell, C. 2006. *ISD from the Ground Up.* 2nd ed. Alexandria, VA: ASTD Press.

International Federation of Biosafety Associations. 2012. IFBA Certification Program Ensuring Quality Biorisk Management through Certification of Professionals. IFBA.

Jensen, E. 2005. *Teaching with the Brain in Mind.* 2nd ed. Alexandria, VA: Association for Supervision and Curriculum Development (ASCD).

Jensen, E. 2008. *Brain-Based Learning the New Paradigm of Teaching.* 2nd ed. Thousand Oaks, CA: Corwin Press.

Kahn, L.H. 2012. Can Biosecurity Be Embedded into the Culture of the Life Sciences? *Biosecurity and Bioterrorism: Biodefense Strategy, Practice, and Science*, 10: 241–246.

Kaufman, S.G., and R. Berkelman. 2007. Biosafety "Behavioral-Based" Training for High Containment Laboratories: Bringing Theory into Practice for Biosafety Training. *Applied Biosafety: Journal of the American Biological Safety Association*, 12: 178–184.

Kirkpatrick, D.L. 2007. *The Four Levels of Evaluation: Measurement and Evaluation.* Alexandria, VA: American Society for Training & Development.

Kolb, D.A. 1984. *Experiential Learning: Experience as the Source of Learning and Development.* Englewood Cliffs, NJ: Prentice-Hall.

Koo, D., and K. Miner. 2010. Outcome-Based Workforce Development and Education in Public Health. *Annual Review of Public Health*, 31: 253–269.

Lee, A., C. Dennis, and P. Campbell. 2007. Nature's Guide for Mentors. *Nature*, 447: 791–797.

Marsh, H.W. 1980. The Influence of Student, Course, and Instructor Characteristics in Evaluations of University Teaching. *American Educational Research Journal*, 17: 219–237.

Maslow, A.H. 1943. A Theory of Human Motivation. *Psychological Review*, 50: 370–396.

Merens, A., J.D. Cavallo, F. Thibault, F. Salicis, J.F. Munoz, R. Courcol, and P. Binder. 2012. Assessment of the Bio-Preparedness and of the Training of the French Hospital Laboratories in the Event of Biological Threat. *Euro Surveillance: Bulletin Europeen sur les Maladies Transmissibles* (European Communicable Disease Bulletin), 17.

Mtui, G.Y.S. 2011. Status of Biotechnology in Eastern and Central Africa. *Biotechnology and Molecular Biology Review*, 6: 183–198.

Nasim, S., A. Shahid, M.A. Mustufa, G.M. Arain, G. Ali, I. Taseer, K.L. Talreja, R. Firdous, R. Iqbal, S.A. Siddique, S. Naz, and T. Akhter. 2012. Biosafety Perspective of Clinical Laboratory Workers: A Profile of Pakistan. *Journal of Infection in Developing Countries*, 6: 611–619.

National Research Council. 2000. *How People Learn: Brain, Mind, Experience, and School: Expanded Edition*. National Academies Press.

National Research Council. 2011. *Research in the Life Sciences with Dual Use Potential: An International Faculty Development Project on Education about the Responsible Conduct of Science*. Washington, DC: National Academies Press.

National Research Council. 2013. *Developing Capacities for Teaching Responsible Science in the MENA Region: Refashioning Scientific Dialogue*. Washington, DC: National Academies Press.

Raffini, J. 1993. *Winners without Losers: Structures and Strategies for Increasing Student Motivation to Learn*. Boston, MA: Allyn and Bacon.

Rengarajan, K. 2010. Development of a Tool to Perform Systematic Gap Analysis of Biosafety Program Based on Biorisk Management System (CWA15793: 2008). Emory University, Atlanta, GA.

Republic of Kazakhstan. 2012a. Sanitary and Hygiene Requirements to the Health Facilities.

Republic of Kazakhstan. 2012b. Sanitary and Hygiene Requirements to the Laboratories.

Risi, G.F., M.E. Bloom, N.P. Hoe, T. Arminio, P. Carlson, T. Powers, H. Feldmann, and D. Wilson. 2010. Preparing a Community Hospital to Manage Work-Related Exposures to Infectious Agents in BioSafety Level 3 and 4 Laboratories. *Emerging Infectious Diseases*, 16: 373–378.

Rusly, N.S., L. Amin, and Z.A. Sainol. 2011. The Need for Biosafety Education in Malaysia. *Procedia Social and Behavioral Sciences*, 15: 3379–3383.

Russian State Committee for Sanitation and Epidemiological Oversight. 1994. 1.2 Epidemiology Safety in Working with Group I and II Pathogenicity Microorganisms. Moscow.

Salerno, R.M., and J.M. Gaudioso. 2007. *Laboratory Biosecurity Handbook*. Boca Raton, FL: CRC Press.

Sanchez, A.L., M. Canales, L. Enriquez, M.E. Bottazzi, A.A. Zelaya, V.E. Espinoza, and G.A. Fontecha. 2012. A Research Capacity Strengthening Project for Infectious Diseases in Honduras: Experience and Lessons Learned. *Global Health Action*, 6: 21643.

Sundqvist, B., U.A. Bengtsson, H.J. Wisselink, B.P. Peeters, B. van Rotterdam, E. Kampert, S. Bereczky, N.G. Johan Olsson, A. Szekely Bjorndal, S. Zini, S. Allix, and R. Knutsson. 2013. Harmonization of European Laboratory Response Networks by Implementing CWA 15793: Use of a Gap Analysis and an "Insider" Exercise as Tools. *Biosecurity and Bioterrorism: Biodefense Strategy, Practice, and Science*, 11(Suppl. 1): S36–S44.

Tanner, K., and D. Allen. 2004. Approaches to Biology Teaching and Learning: Learning Styles and the Problem of Instructional Selection—Engaging All Students in Science Courses. *Cell Biology Education*, 3: 197–201.

Thorndike, E.L. 1932. *The Fundamentals of Learning*. New York: Teachers College Press.

Turegeldiyeva, D. 2014. *Biosafety Training in Kazakhstan*, ed. L. Grainger. Email Correspondence ed.

US Centers for Disease Contol and Prevention. 2011. Centers for Disease Control and Prevention Morbidity and Mortality Weekly Report 60. Guidelines for Biosafety Laboratory Competency. http://www.cdc.gov/mmwr/preview/mmwrhtml/su6002a1.htm.

US Department of Health and Human Services. 2007. *Proposed Framework for the Oversight of Dual Use Life Sciences Research: Strategies for Minimizing the Potential Misuse of Research Information.* National Institutes of Health, National Security Advisory Board for Biosecurity.

US Department of Health and Human Services. 2009. *Biosafety in Microbiological and Biomedical Laboratories.* 5th ed.

USSR Ministry of Health. 1978. Instruction on Regime of Control of Epidemics while Working with Materials Infected or Suspected to be Infected with Causative Agents of Infectious Diseases of I-II Groups.

USSR Ministry of Health. 1979. Concerning Rules of Registration, Containment, Handling and Transfer of Pathogenic Bacteria, Viruses, Rikketsia, Fungi, Protozoa and Others, also Bacterial Toxins and Poisons of Biological Origin.

Walker, W.O., P.C. Kelly, and R.F. Hume. 2002. Mentoring for the New Millenium. Medical Education Online 7.

Wang, X. 2004. Challenges and Dilemmas in Developing China's National Biosafety Framework. *Journal of World Trade*, 38: 899–913.

World Health Organization. 2006. *Biorisk Management: Laboratory Biosecurity Guidance.*

7 Operations and Maintenance Concepts

William Pinard, Stefan Breitenbaumer, and Daniel Kümin

ABSTRACT

The goal of this chapter is to outline the necessary elements of a proper maintenance system. While the focus appears to be on maintenance, many of the activities undertaken are routine operational activities, thus combining the two topics as a seamless activity. This chapter will provide an overview of the types of maintenance, explain the impact the lab environment can have on maintenance, and the impact that the maintenance can have on the lab environment. It will discuss maintenance documentation, frequency, and roles and responsibilities.

INTRODUCTION

Historically, the requirements for keeping a new building in peak working condition were often not taken into account during the design and construction of the building. This resulted in the deterioration of the building's integrity, and limited the ability of the organization to confidently perform its mission. Maintenance programs focused primarily on fixing equipment and building parts that were broken (Sullivan et al. 2010). Not surprisingly, it was often difficult to anticipate maintenance costs, and to maintain efficient building operations.

These issues are compounded when trying to operate and maintain a biological laboratory. Beyond the usual maintenance requirements, the maintenance plan in a bioscience facility must include unique elements, such as limited access to the equipment and working in areas where there are potentially dangerous pathogens. Traditionally, maintenance plans have addressed this by developing separate plans for each of the biosafety levels. As a result, the facility strategy was a combination of many different—often independent and occasionally conflicting—maintenance plans.

This chapter introduces a risk-based approach for developing a comprehensive maintenance system. This system utilizes the principles of biorisk management discussed throughout this book, which includes risk assessments, implementation of proper mitigation strategies such as practices and procedures or training, and the development of performance metrics. This operation and maintenance approach takes into account the spectrum of risks involved in working in a biological laboratory, as well as the level of disruption expected while completing repairs. The chapter introduces the four primary maintenance strategies: reactive, preventive, predictive, and reliability-centered maintenance. It discusses the personnel responsible for various aspects of maintenance, and concludes with a risk- and disruption-based approach. While the disruptive effects maintenance has on the facility may not directly impact the biological risk, it is still a critical component of the overall facility risk, and management must account for these issues. Properly planning for a disruption will allow for less impact on the mission of the facility, and enable the facility to continue to meet its production, diagnostic, research, or other needs.

Some maintenance strategies will be more effective in certain laboratory settings than others. For instance, an inoperable laboratory because of maintenance can be either acceptable with few consequences, such as in a research laboratory, or of high consequence, such as in a diagnostic laboratory, where the mission does not allow for a shutdown without careful planning. These issues need to be considered when developing a maintenance plan.

FOOT-AND-MOUTH DISEASE VIRUS RELEASE

Operations and maintenance have been shown to be critical in preventing the release of biological agents. One example occurred in September 2007, when an outbreak of foot-and-mouth disease (FMD) virus was identified on two farms near Pirbright, Surrey, England. The outbreak was traced back to a nearby facility, the Institute for Animal Health and Merial Animal Health Limited (Spratt 2007).

Although an independent investigation did not conclusively determine the root cause, several possible causes were identified. The report, known as the *Independent Review of the Safety of UK Facilities Handling Foot-and-Mouth Disease Virus*, also known as the Spratt Report, focused on the integrity of the effluent pipes that led from the facilities to the final waste treatment. Inspectors noted that the pipes were not frequently inspected, and proper funding for maintaining and replacing the old system was not available. The effluent system itself was not even well understood: both its composition and age were not known (Spratt 2007).

A full survey following the outbreak concluded that the effluent treatment pipes did not maintain containment. The final report recommended that "as a matter of urgency, [Department for Environment, Food and Rural Affairs] should require that actions are taken to ensure the effluent drainage system at the Pirbright facility is fully contained and its continuing integrity confirmed by regular inspections. In the interim, we advise that work with the infectious virus should only be allowed if effluent released into the pipes has first been completely inactivated" (Spratt 2007). A separate report by the UK Health and Safety Executive concluded that "weaknesses were identified in the containment standard of the effluent drains across the Pirbright site. These included displaced joints, cracks, debris build-up and tree root ingress. For a biosecurity-critical system, record keeping, maintenance, and inspection regimes were considered inadequate" (Health and Safety Executive 2007).

By properly maintaining the facility and integrating an operation and maintenance program under the umbrella of a biorisk management system, facility managers could identify and promptly repair breaches in the system. An outbreak of FMD virus in the UK in 2001 resulted in 4 million animals slaughtered, an economic loss of $11.7 billion (Pendel et al. 2007), public outrage, and a rise in depression and suicides among farmers (Knight-Jones and Ruston 2013). Clearly, more comprehensive measures should have been taken to prevent a recurrence of this level of economic and personal devastation. The risk assessment should have identified possible leakage through the effluent pipes, and the operations and maintenance program should have developed a plan to mitigate this risk.

MAINTENANCE STRATEGIES

OVERVIEW

Any piece of equipment has a certain likelihood of failure. Depending on the type of equipment, it may have a failure rate that is highest immediately after purchase because of manufacturing defects, or late in its designed life span. Between these two peaks lies an area of low likelihood of failure (Sullivan et al. 2010; Sondalini 2004). Other equipment's failure curves will have a high early onset mortality followed by a stabilized likelihood of failure, minimal early onset mortality with rapid wear-out, constant failure rate over time, rapid end-of-life increase in failure, or steadily increasing likelihood of failure. See Figure 7.1 for illustration of these failure curves. The goal of any maintenance strategy is to extend the low likelihood of failure period for as long as possible.

There are four primary maintenance strategies, each of which has its own advantages and disadvantages. Depending on the situation, it may be best to utilize a variety of strategies. This chapter discusses four maintenance strategies: reactive, preventive, predictive, and reliability-centered maintenance. The reliability-centered

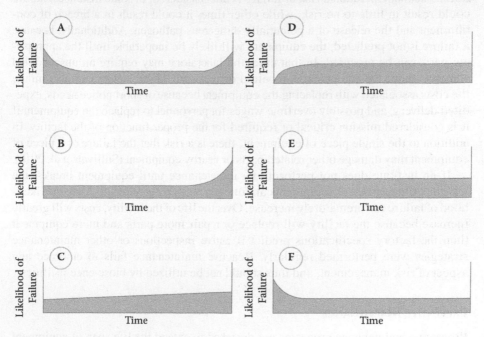

FIGURE 7.1 Failure curves. Failure curves A–C indicate failures based on time-dependent factors. Failure curves D–F indicate failures independent of time.

maintenance strategy was developed by the airline industry in the 1970s to create higher standards for maintenance requirements (Nowlan and Heap 1978); it is a combination of the other three strategies, incorporating the advantages of multiple strategies while minimizing the disadvantages.

While this chapter describes all four maintenance strategies, it recommends reliability-centered maintenance as the preferred method. The other three strategies are described to highlight their individual insufficiencies, and to demonstrate how they can fit in as a part of a combined maintenance strategy.

REACTIVE MAINTENANCE (EVENT-ORIENTED MAINTENANCE)

From a biorisk management perspective, reactive maintenance lacks critical components to adequately address the needs of a risk-based system. Reactive maintenance historically has been the most commonly used maintenance strategy (Sullivan et al. 2010). It is often referred to as the "run until it breaks" strategy. The International Facility Management Association (2014) describes this strategy as "corrective action taken upon failure or obvious threat of failure." In theory, this can be an attractive method. Early in the operation of the building, there is generally little to no cost for maintenance or loss of productivity because of nonfunctioning equipment. Reactive maintenance also maximizes the use of any single part since it is used for its complete life span.

While on the surface this seems to be an effective strategy, using only reactive maintenance can lead to unpredictable failures. It is a system that relies on failure to initiate action. Operational risk of failure is not considered. In some instances, failure could result in little to no risk, while other times it could result in a breach of containment and the release of a potentially dangerous pathogen. Additionally, because a failure is not predicted, the equipment will likely be inoperable until the appropriate parts can be procured. In that case, the laboratory may require an unscheduled shutdown until the equipment is repaired or replaced. These factors can also inflate the costs associated with replacing the equipment because of short notice needs, expedited delivery, and possibly overtime wages for personnel to replace the equipment if it is considered mission critical or required for the proper function of the facility. In addition to the single piece of equipment, there is a risk that the failure of a piece of equipment may damage other related parts or nearby equipment (Sullivan et al. 2010).

If an institute does not perform any maintenance until equipment breaks, the length of time a piece of equipment can function properly is reduced. Also, the likelihood of failure can prematurely increase. Over the life of the facility, costs will greatly increase because the facility will replace or repair more parts and more equipment than the factory specifications predict if active inspections or other maintenance strategies were performed regularly. Reactive maintenance fails to embrace any aspect of risk management, and thus should not be utilized by bioscience facilities.

PREVENTIVE MAINTENANCE

Preventive maintenance programs are designed to extend the life span of equipment and to increase reliability. Preventive maintenance is "the periodic and planned actions taken to maintain a piece of equipment within design operating conditions and extend its life and performed before equipment failure or to prevent equipment failure" (International Facility Management Association 2014). In other words, it is maintenance that is performed according to a predetermined time interval, number of utilizations, or total hours used. This strategy replaces parts before they break. Unfortunately, in most facilities, time-based failures represent a small percentage of all failures (Sondalini 2004), making this strategy inappropriate for a biorisk management system because it does not take into account the likelihood and consequence of failure in a specific situation.

There are many reasons to implement a preventive maintenance plan. Setting up a maintenance schedule allows for greater predictability and flexibility in determining when the equipment is inoperable. Laboratory shutdowns can be planned in advance. It will also expand the life span of the equipment while reducing the likelihood of the equipment failing (Sullivan et al. 2010). Having a predetermined plan allows for predictable costs and maintenance schedules by arranging for the parts and labor ahead of time.

While preventive maintenance lowers the likelihood of failure, it does not eliminate the possibility of catastrophic failure (Sullivan et al. 2010). A preventive maintenance plan does not consider the current condition of the parts, and if the parts are not replaced in a timely manner, the parts have a higher chance of failing. Other disadvantages of using a preventive plan are that the parts are not used for their full life span, and that different parts in the same piece of equipment can have highly variable life spans and maintenance needs. While long time intervals can increase the likelihood of failure, too short of a time interval will result in higher costs and more inoperable time. The development of maintenance schedules through preventive maintenance will consider the generic risk of failure by using factory-provided schedules, specifications, and quality testing data. This makes preventive maintenance more appropriate for use in a biorisk management system than reactive maintenance, but still inadequate because the actual risk based on the condition of the equipment is not taken into consideration.

Many of the disadvantages will become issues if the preventive plan is not executed properly. When the schedule is properly designed, a preventive maintenance strategy can decrease the overall cost of maintenance in the facility by an average of 12–18% over a strictly reactive strategy (Sullivan et al. 2010). This savings is difficult to estimate, as the types of costs are so different between the two strategies. The preventive strategy consists of routine replacement parts and labor costs at a constant but low level, whereas the reactive strategy relies on infrequent replacement but high costs for parts and labor.

However, if the equipment is not used in the ideal conditions established by the manufacturer, then the condition and wear can change, thus changing the timetable for repair. The facility-specific risks of equipment failure are not specifically built into the system. Although a preventive maintenance system is better than a reactive system, it is not appropriate for applying the risk analysis necessary for a biorisk management system.

PREDICTIVE MAINTENANCE (CONDITIONAL MAINTENANCE)

Predictive maintenance is the use of measurements to quantify the level of degradation within a piece of equipment or system. Predictive maintenance consists of those activities involving continuous or periodic monitoring and diagnosis to forecast component degradation so that as needed, maintenance can be scheduled (International Facility Management Association 2014). The primary difference between preventive and predictive maintenance is how to determine when a piece of equipment needs to be replaced. Both strategies strive to extend the low likelihood of failure period for the equipment by replacing parts that wear out over time. While preventive maintenance is based on predetermined time intervals, predictive maintenance uses various methods to determine equipment fitness. Since predictive maintenance uses more in-depth risk assessment methodologies than either preventive or reactive maintenance to determine repair timelines, it is better suited for use in a biorisk management system.

Predictive maintenance programs can use different methods for measuring and quantifying the level of degradation of specific equipment. One method for determining equipment fitness is called performance measurement, which involves tracking the equipment's ability to function properly, and identifying deviations that could be indicative of the need for calibration or repair. A second method is direct inspection of the individual parts, such as looking for wear and damage to the equipment. As wear increases, management can decide to replace the part before it reaches a critical point of failure (Sullivan et al. 2010).

There are several advantages to a predictive maintenance strategy. Similar to preventive maintenance, there is a reduction in the likelihood of failure, and a lengthening of the lifetime of the equipment. By determining the actual state of degradation, the institute can achieve the full life span of the parts. The more data that the institute can collect, the more predictable the maintenance and the costs can be over time. However, continual monitoring is necessary to acquire all of the requisite data. Such continual monitoring requires substantive human and financial resources. Management must be willing to allocate these resources, establish a continual monitoring culture, and train employees. Often, management is unwilling to make the necessary commitment for a predictive maintenance program. Although the cost savings are not readily apparent, a properly functioning predictive maintenance program can save 8–12% compared to a preventive maintenance program (Sullivan et al. 2010). An additional challenge with predictive maintenance is that scheduling repairs based on the condition of the parts results in frequent and often unscheduled maintenance interventions. As a result, a predictive maintenance program may inhibit the facility's operation more than other maintenance strategies.

By collecting the data and making decisions based on the actual conditions, predictive maintenance addresses many of the questions that an assessment of the risk of equipment failure should ask. For instance, monitoring the changes in a refrigerator's temperature allows for predicting when one of its parts may need to be repaired—before the refrigerator fails. However, a predictive maintenance strategy generally fails to consider the risks of exposure, illness, or release: What happens if the refrigerator fails? Is it acceptable for the refrigerator to fail? These types of questions are not asked in a predictive maintenance system. The value of any single piece of equipment is not evaluated within the larger context.

RELIABILITY-CENTERED MAINTENANCE

The US National Aeronautics and Space Administration (NASA) defines reliability-centered maintenance as "the process that is used to determine the most effective approach to maintenance. It involves identifying actions that, when taken, will reduce the probability of failure and which are the most cost effective. It seeks the optimal mix of Condition-Based Actions [predictive], other Time- or Cycle-Based actions [preventive], or Run-to-Failure [reactive] approach" (International Society of Automotive Engineers 1999). This strategy looks at the holistic system to determine relative risk, recognizing that some equipment is more important than others, and some failures will not compromise safety and security.

Thus far, this chapter has discussed time-dependent types of maintenance. While this holds true for some equipment, there are several other failure curves to consider, as seen in images D–F of Figure 7.1. By analyzing the risks of each piece of equipment and its parts, management can determine if the equipment is mission critical, the relative risk of a specific part failing, and the priority in which it should be repaired or replaced. Many organizations suggest flowcharts and questions for determining this schedule. One such example can be found in the Whole Building Design Guide from the National Institute of Building Sciences (Pride 2010). This guide recommends asking the following questions:

1. What does the system or equipment do? What are the functions?
2. What functional failures are likely to occur?
3. What are the likely consequences of these functional failures?
4. What can be done to reduce the probability of the failure(s), identify the onset of failure(s), or reduce the consequences of the failure(s)?

This strategy allows for the allocation of limited funds and manpower to the most critical maintenance issues. The strengths and weaknesses of a reliability-centered maintenance system are most similar to the predictive maintenance strategy. The advantage that reliability-centered maintenance has over predictive maintenance is the recognition that while monitoring certain equipment is critical, some other equipment needs less surveillance. This is the strongest maintenance strategy for use within a biorisk management system because it considers the actual use of specific equipment within the entire system, and it adjusts maintenance strategies based on these new data points.

An example of a rudimentary reliability-centered maintenance system is automobile maintenance. Different parts of the vehicle require varying degrees of attention. Some parts, such as the headlights, radio, and other nonessential components, can be ignored from a maintenance perspective until they break and need repair. Other components, such as engine oil, should be replaced on a regular basis because it is critical, and the maintenance is inexpensive enough that changing the oil prematurely will have little financial impact. Finally, monitoring the tread on tires regularly maximizes the life span of the tires without sacrificing safety.

There are many advantages to such a strategy. First, because this strategy is a combination of the other three, it is able to capitalize on the benefits of each, creating an efficient maintenance program. By eliminating unnecessary repairs, minimizing short turnaround maintenance, lengthening the time between repairs, and creating an efficiently functioning system, an institute can reduce costs to a fraction of the previous costs for strictly reactive, preventive, or predictive maintenance (Sullivan et al. 2010). Unfortunately, these cost savings are not readily apparent to management, especially because reliability-centered maintenance start-up costs for training, equipment, and planning can be significant.

As this section demonstrates, no one strategy is perfect. Each has disadvantages. The reactive and preventive maintenance strategies do not adequately take risk assessment into account. Predictive maintenance characterizes some, but not all, of the relevant risks. Reliability-centered maintenance can help address the shortcomings of each of the other strategies.

Reliability-centered maintenance utilizes an approach that combines predictive, preventive, and reactive maintenance strategies, and relies on a risk assessment to develop appropriate mitigation measures. In this case, conducting a risk assessment determines the likelihood of failure of any piece of equipment or system and the consequences associated with such a failure. The consequences include those associated with pathogen exposure, economic ramifications, production loss, repair times, and costs. Using this characterization, an evaluation and categorization of the various identified risks should be completed to determine the level of attention each system or piece of equipment will require. By using this method in combination with the AMP model, management can design an operation and maintenance schedule that will support a biorisk management program. The reliability-centered maintenance strategy offers the best opportunity to align the safety, security, and maintenance objectives of a biorisk management system.

DEVELOPING A TIERED MAINTENANCE SYSTEM

SYSTEM LAYOUT

Developing and implementing a facility-wide maintenance system can save a significant amount of time, money, and personnel resources. Thorough preplanning forms the basis of a good maintenance system. This not only includes the planning of the maintenance work to be performed, but also considers what should be done before any maintenance work is completed, thus limiting the number of unnecessary equipment failures or breakages.

Such a system is dependent on comprehensive documentation, which articulates what equipment exists, and the maintenance requirements of each item. Furthermore, all the maintenance work—what piece of equipment went through maintenance, what exactly was done, when it was performed, etc.—must be clearly documented. Also, equipment failures should be documented to help influence future decisions on maintenance or replacement strategies.

As discussed in the previous chapters, an institute should always utilize its physical infrastructure and engineering controls to their full potential. A laboratory should be designed to accommodate the users' needs while streamlining work processes. But the maintenance requirements should be included in the laboratory design. For example, equipment should be placed and used in such a way to minimize the hazards and maximize the efficiency of workflow. Efficient lab layout can discourage lab workers from creating workflow solutions that are counter to the equipment's proper use, which could lead to equipment damage. Such steps will limit damage to the equipment and thus achieve cost savings.

A failure of a piece of equipment in one area of the laboratory may indicate a possibility of a failure elsewhere of similar equipment. If a piece of equipment has frequent failures or problems, it is important to address the issue, notify staff of the potential for this error, and monitor similar equipment for that failure. For instance, if a door hits a piece of equipment every time the door is opened, it will cause damage. This damage could be prevented by attaching rubber barriers on the door or the equipment, or by moving the equipment to a different location. Solutions from other sectors may be adapted to specific needs in a laboratory, such as specialized mounts to secure bench tops, minimizing damage due to stress on fixed points, or the use of hands-free sensors to prevent constant wear from physical touch. Developing tailored solutions to prevent damage in each specific situation will better protect the equipment.

ROLES AND RESPONSIBILITIES

Overview

In order for any maintenance activity to be performed successfully, regardless of the complexity of the task, several roles and responsibilities must be defined. All the responsible people involved in maintenance must coordinate with each other. This section describes the roles required for a maintenance program, and outlines certain responsibilities these people should have.

Executive/Mid-Level Management

Ultimately, the successful operation of any laboratory requires the full support of both executive and mid-level management. Management must acquire and provide the financial means to run a program. Management must also obtain enough financial resources to pay for repairs, spare parts, the exchange of parts or whole machinery, as well as to pay for personnel to perform the maintenance. Without this support, a maintenance program cannot be successful.

Every institution is organized differently, and may have different hierarchical structures. Regardless, effective and efficient communication is a characteristic of any good organization. For instance, the engineering manager must work with laboratory leadership to set timelines for operational stoppages, and establish proper roles and responsibilities within the entire laboratory. It is the responsibility of management to make sure all groups involved communicate with each other regardless of what part of the institution they work for.

Furthermore, management must understand the time requirements for maintenance. This includes not only the time required for the actual maintenance work to be performed, but also the planning beforehand and the necessary tests that must be run following completion of the work. In some cases, training of staff may also be required prior to any maintenance work, and this training may require a significant investment of time and financial resources.

Maintenance Personnel

Clearly, maintenance personnel play the most important role in any maintenance program. They are involved in planning the maintenance activities, undertaking the work, and ensuring that the facility runs within the specified parameters following these activities. In addition to a thorough knowledge of the facility features, maintenance personnel must understand how the different parts act together to achieve the smooth and safe operation of the facility. In other words, the maintenance personnel need to understand how engineering supports all aspects of the biorisk management system.

Maintenance personnel generally have a background in engineering. However, they often lack a familiarity with the biological hazards possibly encountered in a laboratory or a plant room that supports the laboratory. Depending on what maintenance work individuals perform, they may require additional training in basic microbiology to understand the biological risks involved with their work. In a low hazard environment, that training may be rudimentary because the risks posed by organisms in that environment may be easily controlled. However, should maintenance staff be required to enter an extreme hazard environment, that training should be very extensive.

Training of staff is an important factor in any maintenance program. In addition to basic microbiology, maintenance staff should also be trained in the use of certain equipment, such as special tools or personal protective equipment (PPE). They also need to be aware of the context a certain part plays within the whole system, and what the consequences of a failure of that part may be regarding biosafety or biosecurity. If chemicals such as decontaminants are used, maintenance staff should also receive training in chemical safety. In general, training for maintenance staff must be as carefully considered as it is for anyone else who works in the facility.

Depending on the complexity of a facility, it is important to make sure that all necessary trades are covered. Most large facilities employ at least one electrician and one mechanical engineer. In some cases, it may be necessary to employ specialists in HVAC and plumbing/wastewater treatment systems. If the facility is particularly complex, then a member of the staff should have expertise in building management systems. If a facility does not have maintenance personnel on staff, outside contractors can be used. Additional considerations must be made to ensure proper safety training and responsibilities are clearly defined in the business agreement or contract.

Biorisk Management Professionals

The biorisk management professional provides the link between the laboratory and the maintenance personnel. He or she should be able to facilitate communication between scientists and the engineers, and thus make sure both sides understand the purpose of each other's work. Also, the biorisk management professional must ensure that the environment is safe for the maintenance staff to perform their work. Following any maintenance work, the biorisk management professional, together with the responsible engineer, should review the work performed and determine whether the facility can return to normal operations. In order to best perform these tasks, the biorisk management professional should understand basic engineering principles and the specifics of the equipment maintenance strategy and program.

Laboratory Personnel

Laboratory personnel play an important role in any maintenance strategy. Simply by using the laboratory equipment properly—based on recommendations from manufacturers, the engineering staff, and the biorisk management staff—laboratory personnel can drastically reduce the amount of time the facility is out of operation for equipment maintenance. For example, if laboratory personnel use bleach to decontaminate the stainless steel surfaces of a biosafety cabinet, those surfaces will corrode prematurely unless additional cleaning follows decontamination.

Laboratory personnel may also provide valuable support by collecting data about their equipment for predictive maintenance. They can monitor equipment performance, which will help plan maintenance work. Where appropriate, laboratory staff may also handle simple maintenance work to minimize interruptions by the maintenance staff. Finally, staff is responsible for the prompt notification of proper individuals in the event of an equipment problem that they cannot resolve.

Environmental Health and Safety

Most large facilities have an environment, health, and safety (EH&S) group. In a bioscience facility, the health of the maintenance staff is a critical factor that may need to be monitored. For example, it may be necessary to assess the need for engineering staff to be vaccinated prior to undertaking any maintenance work. While it is obvious that laboratory staff undergo preventive medical programs, engineering staff and outside contracting personnel are often overlooked. Some maintenance work may be required to take place during normal operations, and thus engineering staff should be included in established preventive medical programs.

EH&S personnel may also be required to address other hazards that are present in the laboratory in addition to the biohazardous considerations. For example, maintenance workers need to be informed of both the physical and chemical hazards that are present in the laboratory to avoid injury.

MAINTENANCE CONTEXT

Overview

All work involving biological agents should be based on a proper risk assessment. The same holds true for any maintenance work, especially when it is to be performed in a laboratory environment. In addition to the agent-based risks, the institute must evaluate the impact that maintenance requirements will have on facility operations and the scientific program. Here, we outline three levels of hazards that should be incorporated into a risk assessment specific to maintenance systems: low, high, and extreme. While the following section outlines distinct levels, they are only meant to be a guide and a way of thinking about the hazard—not a dogmatic, tiered list.

Low Hazard

Low hazard environments may be characterized as follows:

- Access possible at all times to the laboratory equipment and systems
- No special safety measures
- Maintenance work relatively unhindered by operational environment
- Low constraints for disposal
- Low requirements for preparing equipment and space to be safely accessible
- Both planned and unplanned stand-downs possible

In situations where only low risks are present and where access is possible at all times, the laboratory should rely on a combination of preventive and predictive maintenance strategies. It is always sensible to plan maintenance work, but it may not always be possible. However, if maintenance work is planned in advance and communicated to other involved staff properly, this will most certainly reduce staff stress and help to promote a better working environment.

High Hazard

High hazard environments may be characterized as follows:

- Limited access to laboratory equipment and systems
- Special safety and security measures and considerations
- Maintenance work limited in scope during laboratory operations
- Complex procedures for entry and exit of materials
- Significant requirements for preparing equipment and space to be safely accessible
- Require prior planning for any stand-downs
- Alternate sample processing may be necessary (outsourcing to other laboratories or facilities)

In a high-risk environment, maintenance work should only be performed during predetermined maintenance intervals. Thorough planning of all maintenance work will allow for limited stand-downs or disruptions to the operation of the facility. This is important for several reasons. It may be time-consuming to "take down" or close the facility for maintenance work, and it may be equally time-consuming to return the facility to normal operations. A lengthy laboratory shutdown may make it necessary to implement contingency plans, such as outsourcing certain activities, for laboratory operations.

Extreme Hazard

Extreme hazard environments may be characterized as follows:

- Access strictly limited to laboratory equipment and systems
- Specialized protective suits and procedures

- Maintenance work very limited during laboratory operations
- Complex procedures for entry and exit of materials, may even be impossible to bring certain equipment for maintenance use
- Require prior planning for any stand-downs
- Alternate sample processing may be necessary (outsourcing to other laboratories or facilities)

All maintenance work should be performed during predetermined maintenance intervals only. Only very limited, if any, maintenance work should be done during normal laboratory operations. Most maintenance staff are not trained to enter these facilities or areas during normal operations, and a complete stand-down of the facility is thus required for most, if not all, maintenance work. Because these facilities are often quite complex, it requires considerable time to completely shut down operations, and then return the laboratory to normal operations. Thus, all maintenance work should be planned well in advance. Also, contingency plans for any laboratory work to be performed during maintenance will need to be put in place.

MAINTENANCE CATEGORIES

Overview

As mentioned previously, prior planning of maintenance work helps extend the life span of equipment parts and facilities, limits the amount of facility downtime, and facilitates coordination between laboratory and engineering staff. Similar to the hazard categories, the following maintenance categories are a way to organize the information and should not be taken as absolute levels. The four categories are based on the following requirements:

- Impact on operational safety and security
- Impact on laboratory operation (users)
- Impact on availability of laboratory infrastructure (acceptance of samples)
- Timely influence (expected stand-down times)
- Planning horizon
- Returning to normal operations
- Need for fumigation

Maintenance frequency depends on the type of equipment used, run times of the equipment, mission criticality of the facility, and results of inspections performed on the equipment. When defining maintenance strategies and frequencies, one also needs to consider necessary lead times for certain parts. For example, certain parts may need to be ordered well in advance because they may not be readily available or may be manufactured only following placement of an order. Thus, certain critical parts should be stored on site in order to provide full engineering support to the institution. The maintenance staff should define critical parts for a piece of equipment together with the manufacturer, determine what parts may need to be held in stock on site, and define their use over time. In general, these parts will be used during the

TABLE 7.1
Overview of Maintenance Categories and Their Characteristics

Category	A	B	C	D
	No impact on safety and availability	Impact on safety and availability		
Type of maintenance work	Inspection and small services	Shorter, less extensive activities with maintenance interval ≤ 1 year	Extensive work on important systems with maintenance interval ≥ 1 year	Very extensive work on whole facility
Stand-down time[a]	None	Days to weeks	Several weeks	Months
Planning horizon	None	Months	1 year	Several years
Implementation of maintenance work	Continuously	At predetermined intervals, once every 3–6 months	At predetermined intervals, 1× per year	As needed
Time needed to return to normal operations	N/A	Possibly 1 day	Several days	Time-consuming, functionalities may be newly defined
Analysis possible	Yes	Only if lab still in operation	No	No
Examples	Observe system parameters, refilling working materials, exchange wear parts	Periodic controls, service/repair of lab equipment, cleaning	Maintenance HVAC, ETP,[b] autoclaves, breathing air supply	Rare, larger modifications and renewal

[a] Not including time for prior decontamination.
[b] Effluent treatment plant.

next maintenance activities and be immediately replenished. For an overview of the proposed maintenance categories, see Table 7.1.

Category A: Low to No Impact on Facility Operations

Maintenance category A includes activities that do not impact personnel safety or security, or the availability of the facility. Tasks include inspection and service work that do not require shutting down the facility. Work is usually performed outside of containment, and thus there is no need for fumigation. Simple inspections inside containment may also be performed by users (scientific personnel).

Examples include:

- Observation of facility values
- Performance measurements (e.g., air pressures or refrigerator temperatures)

- Small cleaning and lubrication work
- Exchange of worn parts (only those that may be exchanged during operations)

Category B: Low to Moderate Impact on Facility Operations

Activities performed in category B maintenance work may have an impact on personnel safety and security, and facility availability. This includes mostly laboratory equipment and plant parts that require maintenance intervals of less than one year and thus cannot be maintained during yearly maintenance periods.

Tasks include mostly shorter and less extensive maintenance work on equipment and parts. Typically, several months are needed to plan in advance for category B maintenance. Stand-down time ranges from days to weeks. Activities can be planned for predetermined intervals (every three to six months). Returning to normal operations is not particularly elaborate and may only take one day.

Examples include:

- Maintenance of Class II or III biosafety cabinets
- Periodic controls and performance tests
- Filter tests and exchange
- Repair of lab equipment

Category C: Moderate to High Impact on Facility Operations

Category C includes maintenance work that has a moderate to high impact on safety, security, and facility availability. For this type of category, whole systems need to be turned off, possibly including the fumigation of rooms or filter housings, for example. A thorough decontamination of potentially contaminated parts will be necessary prior to commencement of maintenance work.

Work is performed at predetermined intervals (e.g., once a year). Because these activities can be labor-intensive, it may be necessary to plan for several months or up to one year prior to the commencement of maintenance work. The facility may be shut down (at least partially) for several weeks. Furthermore, returning the facility back to normal operations may take several days. In most cases, systems must undergo a number of tests following maintenance before they can be used by laboratory staff.

Examples include:

- Periodic checks and performance tests (e.g., emergency, fire, air lock controls)
- Maintenance of chemical showers or effluent treatment plant (ETP)
- Maintenance of breathing air system or autoclaves
- Replacement of lab equipment
- Extensive lab cleaning
- Complete HVAC revision

Category D: High to Extreme Impact on Facility Operations

Category D is reserved for extensive work on complete systems or facilities. These activities have a significant impact on the safety, security, and availability of the facility. Decontamination of parts, or even the whole facility, will be necessary prior to commencing maintenance work. Because these activities are very elaborate, they need to be planned well in advance (several years) and performed as required. Returning the facility to normal operations following completion of maintenance work will be time-consuming, as certain functionalities will be newly defined.

Examples include:

- Complete renewal of important systems, such as HVAC.

Unplanned Repairs

Not everything can be planned for; unforeseen events will always occur. Preparing for these is very difficult, but an institute must be ready for any unplanned failures and necessary repairs. Every bioscience facility should have plans in place to address unplanned repairs. A generic plan may include:

- Precise analysis of problem
- Develop hypothesis, search for cause
- Define possible approach
- Risk assessment by biorisk management professionals and engineers
- Decision of biorisk management professional/head engineer regarding further actions
- Solve problem immediately or leave until next maintenance period

Regardless of what plans may be instituted, one crucial question should always be: Has there been or will there be a breach of containment? If a breach of containment is not imminent, the situation can be thoroughly analyzed and the necessary countermeasures defined precisely. In other cases, there may be the need for immediate measures to be implemented. These measures can be predefined based on described emergency scenarios for the institute. Following maintenance, the facility must be returned to normal operations. The facility must clearly define who in management will approve a return to normal operations. Finally, simple repairs are usually performed immediately following consultation with the biorisk management professional. Once again, this shows that close collaboration between the head engineer and the biorisk manager is crucial and will provide for smooth operations of the facility.

IMPLEMENTATION

Maintenance Planning

Thorough planning of the maintenance program is critical and will help to make sure that everything runs smoothly. This can be done in two different ways:

1. Use special maintenance software, which is either commercially available or developed locally using available resources.
2. Use a hardcopy or file card system, including all the necessary information.

Commercial software systems allow you to create maintenance tasks in a way that reflects the maintenance plan of the overall facility. Usually, tasks will be captured with their room number and facility system affiliation. Additional data included are outlined in the following section. Based on recommended maintenance intervals, activities can be planned. These software systems create order forms for each maintenance activity with a due date. These forms can be signed and reported back following completion of work. In this way, all work is accounted for and completely documented.

Any nonelectronic system used may incorporate these same principles. The difficulties lie in time management and scheduling of work. Automatic notifications are not possible. However, a calendar system may work just as well, simply requiring the user to check calendar entries on a regular basis.

LIST OF FACILITY SYSTEMS

Every facility should have a comprehensive, up-to-date list of plant systems. This list should include all the equipment in the plant, and pertinent information about that equipment and its spare parts. This list can be administratively or electronically controlled. The data on this list may include the following:

- Precise description of part (name, article number, manufacturer, unique identifier, etc.)
- Purchase (supplier, alternative supplier, delivery time)
- Life span
- Recommendations for storage
- Necessary documentation for installation and removal
- Location

MAINTENANCE ACTIVITIES

To ensure that maintenance is performed completely, the institute should document all necessary and recommended maintenance information for every system (including components and inspection points). This information may then be organized based on maintenance intervals and who can perform them. In effect, the facility should have an easily accessible resource that contains all the required information for any given maintenance object on any given date. In addition, this document should include legal requirements and possible hazards. Table 7.2 is an example of such a list.

ARRANGEMENT INTO MAINTENANCE CATEGORIES

Usually, the head engineer of a facility is tasked with separating maintenance activities into specific maintenance categories. For this, the engineer may need to enlist help from the design and construction team of the facility, manufacturers, and the biorisk management staff. Generally, the aim is to perform as many activities as possible outside containment or in areas that have been approved as safe following decontamination. Maintenance activities should fall into normally scheduled maintenance periods, and meet the desired maintenance strategy (see Table 7.2). Clearly, not all maintenance requirements fall under the same category, nor does every piece of equipment have the same legal or engineering requirements. Thus, a major

TABLE 7.2
Possible Maintenance Activity Overview

This table shows an exemplary overview of maintenance activities and how they may be set up. In its electronic form, further information, such as legal requirements, may be linked to the document, thus incorporating all necessary information in one place.

System Number	Name	Regulatory Requirements	Activities	Interval	Realization Internal/ External	Time Required	Maintenance Category
J01	Lift	Yes	Periodical service	3 monthly	Elevator, Inc.	1 day	—
I59	Autoclave BL-43	No	Cleaning of seals	Weekly	Internal	1 hour	A
I59	Autoclave BL-43	No	Functional checks	Before every use	Internal	5 minutes	A
I59	Autoclave BL-43	No	Periodical service	Yearly	Sterile Ltd.	5 day	C
I66	BSC Class III	Yes	Periodical service	Half yearly	Glovebox, Inc.	3 day	B

challenge is to correctly define the right category for every piece of equipment. The following criteria may help in arranging equipment into maintenance categories:

- Legal requirements
- National or international standards
- Good practices
- Manufacturer's recommendations (maintenance intervals, parts to be changed, life span, etc.)
- Maintenance/repair history of a piece of equipment
- Required availability for scientific staff
- Stress factors (temperature changes, mechanical stress, liquids, etc.)
- Run times
- Relationship with/connection to other plant systems (e.g., failure of one piece of equipment may cause damage to another)

CONCLUSION

Explicitly integrating an operations and maintenance strategy into a biorisk management system will significantly reduce the risk of an accidental or intentional release of a dangerous organism. As was shown by the FMD example, lack of routine maintenance resulted in a release of FMD into an environment where the virus did not occur naturally. Proper maintenance and repair of the effluent system could have safeguarded the local environment, livestock, and farmers. If the risk assessment had substantively considered the operations and maintenance of the facility's effluent system, that incident likely could have been avoided.

Using reliability-centered maintenance and properly designing a maintenance plan will reduce the likelihood of equipment failure throughout a facility while minimizing the long-term costs of the maintenance system. Reliability-centered maintenance will also minimize the impact on the operations of the facility caused by unexpected stand-down times for repairing critical equipment. Substantive organizational planning by all those involved, including maintenance personnel, biorisk management professionals, environmental health and safety professionals, and management, will ensure that the workers who conduct the maintenance, the laboratory workers, and everyone else in the facility will remain safe. A robust maintenance program will, in turn, reinforce the technical systems that support the other elements of a biorisk management program discussed in this book. Without a reliability-centered maintenance program, the laboratory is likely to spend larger amounts of money on a less reliable strategy, resulting in increased risk to all personnel and the surrounding environment.

REFERENCES

Health and Safety Executive. 2007. *Final Report on Potential Breaches of Biosecurity at the Pirbright Site.* London: Health and Safety Executive, p. 30. http://www.hse.gov.uk/news/archive/07aug/finalreport.pdf.

International Facility Management Association. 2014. Facility Maintenance Glossary. http://community.ifma.org/fmpedia/w/fmpedia/default.aspx (accessed July 2014).

International Society of Automotive Engineers. 1999. *Evaluation Criteria for Reliability-Centered Maintenance (RCM) Processes.*

Knight-Jones, T.J.D., and J. Ruston. 2013. The Economic Impacts of Foot and Mouth Disease—What Are They, How Big Are They and Where Do They Occur? *Preventive Veterinary Medicine*, 112(3–4): 161–173.

Nowlan, F.S., and H.F. Heap. 1978. *Reliability-Centered Maintenance.* San Francisco: Dolby Access Press.

Pendel, D.L., J. Leatherman, T.C. Schroeder, and G.S. Alward. 2007. The Economic Impacts of a Foot-and-Mouth Disease Outbreak: A Regional Analysis. Paper presented at the Western Agricultural Economics Association Annual Meeting, Portland, Oregon, July 29–August 1, p. 4.

Pride, A. 2010. *Reliability-Centered Maintenance (RCM).* Washington, DC: National Institute of Building Sciences. http://www.wbdg.org/resources/rcm.php?rom.

Sondalini, M. 2004. How to Use Condition Based Maintenance Strategy for Equipment Failure Prevention. *Lifetime Reliability.* http://www.lifetime-reliability.com/free-articles/maintenance-management/condition-based-maintenance.html.

Spratt, B.G. 2007. *Independent Review of the Safety of UK Facilities Handling Foot-and-Mouth Disease Virus.* London: Department for Environment, Food and Rural Affairs. http://archive.defra.gov.uk/foodfarm/farmanimal/diseases/atoz/fmd/documents/spratt_final.pdf.

Sullivan, G.P., R. Pugh, A.P. Melendez, and W.D. Hunt. 2010. *Operations and Maintenance Best Practices: A Guide to Achieving Operational Efficiency.* Washington, DC: Pacific Northwest National Laboratory for US Department of Energy, p. 5.1. http://www1.eere.energy.gov/femp/pdfs/omguide_complete.pdf.

ADDITIONAL REFERENCES CONSULTED

Deutsches Institut für Normierung e.V. 2010. DIN EN 13306: Instandhaltung—Begriffe der Instandhaltung.

Deutsches Institut für Normierung e.V. 2012. DIN 31051: Grundlagen der Instandhaltung.

8 Evaluating Biorisk Management Performance

LouAnn Burnett and Patricia Olinger

ABSTRACT

A biorisk management system functions to control risks associated with handling, storage, and disposal of biological agents and toxins. The AMP (assessment, mitigation, and performance) model for biorisk management requires that control measures (mitigation) be based on a substantive risk assessment (assessment) and also that the effectiveness and suitability of the control measures be evaluated (performance). Although performance is generally evaluated after mitigation is implemented, designing appropriate performance indicators must be part of the planning process. Common strategies used currently to measure performance often rely on failure data and may benefit by utilizing additional indicators that measure more proactive activities and outcomes. Using the results from performance measurement is imperative to make effective changes or improvements to the system, and a continually improving system is a hallmark of an effective biorisk management program.

INTRODUCTION

The English word *performance* first came into use in the 1530s, and in that early context, it meant to carry out or see through a promise or duty (Harper 2014). This original intent of the word applies well to the concept of biorisk management performance—which is carrying out the duty of biorisk management. A management system is the framework of processes and procedures used to ensure that an organization can reliably fulfill all tasks required to achieve its objectives (Wikipedia 2013). Performance, then, is a hallmark of a management *system*—assuring, in a cyclical and ongoing fashion, that implemented measures actually address the objectives established.

As has been seen throughout this book, a biorisk management system can be divided into three primary functions—assessment, mitigation, and performance (AMP). In the context of the AMP model, biorisk management performance measurements provide evidence that the organization can reliably fulfill all the tasks required to conduct substantive, transparent, and comprehensive biorisk assessments and to implement appropriate biorisk mitigation strategies to reduce or eliminate the identified risks.

Table 8.1 lists common performance tools often cited to measure biorisk management performance. These tools, while providing some value in measuring performance, are limited—especially when used in isolation and in a reactive manner. Although important to ongoing function and continual biorisk management improvement, these limited provisions rarely appear proactively in current biorisk management plans, and if they do, they are almost never fully executed. This lack of continual attention and follow-through leaves biorisk management efforts vulnerable to criticism, pushback, deterioration, and at worst, unforeseen failure.

The process described in this chapter advocates more thoughtful performance measurements using actual data to support and develop programs that reduce risk. This, in turn, builds confidence and promotes buy-in at all levels. A carefully planned and fully executed biorisk management system that uses the AMP model provides workers, managers, biorisk management advisors, and institution leadership with documented, risk-based, cost-effective, and locally relevant risk reduction strategies.

Making the shift to an effective, performance-based biosafety and biosecurity program requires three primary actions:

1. Establish performance indicators during the planning stage—concurrently with the process of setting goals and objectives and assigning roles and responsibilities for each new project or program.
2. Involve persons at every level and function of an organization to conduct biorisk management performance measurements.
3. Evaluate and use existing data to create and expand performance measurements.

Controlling "risks associated with the handling or storage or disposal of biological agents and toxins in laboratories or facilities" defines the *purpose* of a biorisk management system, according to Section 1 of the international consensus document CWA 15793:2011 (European Committee for Standardization 2011). Section 4.5.1 of CWA 15793 describes *performance measurement* and analysis of data, stating: "The organization shall ensure that appropriate data are determined, collected and analysed to assess the suitability and effectiveness of the biorisk management system and to evaluate where continual improvement of the system can be made." Using these two provisions together, measurement and evaluation of a biorisk management system should then focus on the activities associated with reliable control of the risks associated with the handling, storage, or disposal of biological agents and toxins, and the extent to which that control has been put into place and can be demonstrated.

THE IMPORTANCE OF PERFORMANCE TO BIORISK MANAGEMENT

Measurement and evaluation of biorisk management performance serves several functions. First, and perhaps most obvious, performance should evaluate that existing biorisk assessments are accurate and that biorisk mitigation measures are suitable to control, reduce, or eliminate the identified risks. Well-designed performance measurements can provide an indication of where some biorisk management strategies are not operating as intended or are deteriorating over time. If metrics are designed

TABLE 8.1

Tools That Have Been Cited and Utilized as Performance Indicators for Biorisk Management

Tool	Description	Advantages as Performance Tool	Disadvantages as Performance Tool
Audits	Planned and documented activity by qualified personnel to determine by investigation, examination, or evaluation of objective evidence the adequacy or compliance with established procedures or documents and the effectiveness of implementation.	• Comprehensive system review.	• Too infrequent to identify system deterioration. • Focus on compliance rather than ensuring that the system is delivering desired outcomes (OECD Environment Directorate 2008). • Complex. • Time-consuming.
Inspections	Organized examination or formal evaluation exercise. Results are usually compared to specified requirements and standards. Inspections can be part, but not all, of an audit.	• Immediate assessment. • Focused on a specific aspect.	• Generally focused on specific aspects of worker safety rather than on overall system function (OECD Environment Directorate 2008). • Planned inspections may not reflect actual conditions. • Unplanned inspections can cause unusual responses due to stress and surprise.
Questionnaires	Instrument consisting of a series of questions and other prompts for purposes of gathering information from respondents.	• Standardized responses help in compilation and comparison of data. • Inexpensive. • Perception has been demonstrated to be a particularly effective indirect measurement of system performance (Choudry et al. 2007; DeJoy et al. 2004; Gershon et al. 2000; Turnberg and Daniell 2008).	• Not always designed for statistical analysis of results. • Low return rates. • Questions can be poorly designed. • Generally completed by persons who are either very pleased or very displeased.

continued

TABLE 8.1 (continued)
Tools That Have Been Cited and Utilized as Performance Indicators for Biorisk Management

Tool	Description	Advantages as Performance Tool	Disadvantages as Performance Tool
Interviews	An interview is a conversation between two or more people where questions are asked by the interviewer to elicit facts or statements from the interviewee.	• Focused on individual interviewee and his or her opinion. • Interviewer can introduce flexibility into conversation. • Can derive depth of detail from the interviewee.	• Time-consuming and resource-intensive. • Very difficult for interviewer to assure that he or she has captured all the information. • Quantitative analysis is difficult. • Difficult to compare populations.
Training evaluations	Training evaluations are most commonly known as questionnaires for students to provide immediate feedback about the instruction of a course or training event. Three additional levels of training evaluation have been identified by Kirkpatrick and Kirkpatrick (1994) that involve other mechanisms than a written questionnaire.	• For level 1 and level 2 evaluations, an immediate assessment of student satisfaction and initial instructor assessment of learning is generated (Kirkpatrick and Kirkpatrick 1994). • If all levels of the Kirkpatrick model are used, satisfaction, learning, behavior changes, and organizational changes are all documented.	• Behavior and organizational changes are best measured 3–6 months beyond training (Kirkpatrick and Kirkpatrick 1994). Due to this time gap, this type of evaluation often does not occur.
Incident reports	A form completed to record details of an unusual event that occurs at a facility. The purpose is to document exact details while they are still fresh in the minds of those involved.	• The report of a defined incident indicates a failure in the system. • An accurate report can clarify liability issues stemming from the incident. • Near-miss reports can show problems with the system before failure occurs. • Investigation of the cause of the incident can help pinpoint areas for closer review.	• Measuring the numbers of reports may not discern the difference between successful reporting and incident occurrence (Choudry et al. 2007; Glendon and Litherland 2001). • Using declining incidents may be a disincentive to reporting. • Success of biorisk management cannot be determined by levels of system failure. • Reports can be of dubious accuracy—especially if reportees expect punitive action.

to provide an early warning, this will allow preventive action to be put in place before a damaging incident occurs—a release, injury, or mishandling of biological materials. The reliance on more proactive performance measurements, rather than on failure data, avoids discovering weaknesses in the system through costly incidents (Choudry et al. 2007; Fernandez-Muniz et al. 2007; Glendon and Litherland 2001; OECD Environment Directorate 2008).

In addition to affirming the control, reduction, or elimination of biorisks, measurement and evaluation of biorisk management performance provide an opportunity to further refine and improve the system by revising goals and objectives, and allocating resources where they will provide the most benefit. Identifying obsolete or ineffective control measures allows reallocation of resources to other strategies. Ongoing system performance assessments allow for continuous improvement of the biorisk management system to identify and address new knowledge and changing needs.

Although performance measurements have not historically been established for biosafety and biosecurity, other industries routinely use safety performance measures (see Chapter 1). Safety science literature contains reports and studies on what types of performance measurements are most indicative of safety performance. Organizations that measure performance for safety or security report that due to the focus on reliable risk management, they have protected their reputation and been able to demonstrate documented suitability of their risk control systems. In addition, these organizations have found that they were better able to use information that was being collected for other purposes (OECD Environment Directorate 2008). Industries reporting these benefits include the construction industry (Choudry et al. 2007), healthcare (DeVries et al. 1991), aviation (Eherts 2008), chemical industrial areas (Reniers et al. 2009), and university and college laboratories (Wu et al. 2007), among others.

One of the most comprehensive documents that describes the value and thought behind safety performance indicators was published by the Organization for Economic Cooperation and Development (OECD) in 2008: *Guidance on Developing Safety Performance Indicators Related to Chemical Accident Prevention, Preparedness and Response* (OECD Environment Directorate 2008). Although this international best practice document describes safety performance indicators for the chemical industry, these processes can easily be transferred and utilized for biorisk management systems. Figure 8.1 outlines the steps described in the OECD publication to systematically develop, implement, and evaluate safety performance indicators.

ESTABLISHING BIORISK MANAGEMENT PERFORMANCE MEASUREMENTS

To follow the steps in Figure 8.1, some basic definitions and processes are necessary. This section identifies the concepts, terms, and basic examples—primarily derived from the OECD safety performance indicator document (OECD Environment Directorate 2008), with added illustrations relevant to biorisk management. The following sections will use these steps in more specific and detailed biorisk management examples.

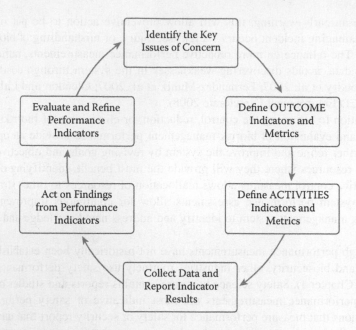

FIGURE 8.1 Steps for implementing safety performance indicators.

Because establishing performance measurements should occur during the planning phase, the lead person responsible for setting the biorisk management system goals and objectives should also be responsible for assuring that performance measurements have been clearly defined and identified, and the roles and responsibilities for gathering those measurements have been appropriately assigned to knowledgeable personnel. However, as is the case for nearly all aspects of biorisk management, appropriate measurements are best established with input and support from all levels—management, technical experts, and employees with hands-on knowledge of the affected work.

Step 1: Identify the Key Issues of Concern

Choosing What to Measure

Measuring performance is most critical where:

- Risk is the greatest
- Controls are the most vulnerable to deterioration over time
- Safety or security problems have occurred in the past
- Gaps in safety or security have been identified
- Newer, more untested mitigation strategies are being implemented

Although tempting, performance indicators should not be chosen merely because they are easiest to measure (e.g., counting the number of days without an incident) or because they make the organization appear successful. Similarly, performance indicators should be developed for what *should* be measured, not for what *can* be measured. Limiting the number of performance evaluations to those of the highest priority allows resources to be focused and used more effectively. Over time, more indicators may be implemented, as earlier goals are achieved or as measurements become more routine.

STEPS 2 AND 3: DEFINE OUTCOME AND ACTIVITY INDICATORS AND METRICS

Basic Definitions for Performance Indicators

A performance indicator has two parts—a definition and a metric. The definition describes what success would look like. The metric describes the measurement to be taken to detect that success. For example, in training, a defined performance indicator might be: Are students able to recall key messages from the training? One metric to measure success could be the score from a quiz given after the training to test student recall of the key messages.

Metrics do not always result in actual numbers. The simple binary yes or no serves an extremely valuable purpose in performance measurements in a system. There are (at least) three categories of metrics:

1. Descriptive metrics: A condition measured at a certain point or time, including, but not limited to, sums, percentages, and composites. Yes or no is a descriptive metric. A descriptive metric for the training indicator given above is the average score from students who took the quiz.
2. Threshold metrics: Compares descriptive metrics to threshold or tolerances, where they exist. A threshold metric could be the number of students scoring a 90% or higher on the quiz.
3. Trended metrics: Changes in descriptive metrics over time. Changes in training recall could be measured using a trended metric of the score of the same students on the quiz (1) immediately following the training, (2) two weeks following the training, and (3) three months after the training.

All of the example metrics provide a measurement for the indicator "Are students able to recall key messages from the training?" However, each metric demonstrates a different way of looking at that indicator. Choosing a metric is as important as defining the indicator.

Types of Performance Indicators

The OECD publication describes two types of performance indicators—outcome and activity indicators. Outcome indicators are designed to define and measure the *extent* to which the system component achieves the desired results (e.g., meeting the *outcome*). Activities indicators measure whether the organization demonstrates the *intent* to develop and implement necessary policies, programs, and procedures (e.g., implementing activities) to support the system components that are in place. These two types of indicators are termed leading and lagging in other safety science literature (Jovasevic-Stojanovic and Stojanovic 2009; Neal and Griffin 2006). However, the terminology is less important than the concept that certain activities must be in place (activities/leading) for desired outcomes to be measured and ultimately achieved (outcome/lagging).

A difficulty in establishing biorisk management performance measurements is that biorisk management is, for the most part, a large and intangible concept, and success is measured by the absence of incidents or problems—however, there is a distinct difference in the probability of future incidents between a lucky lab and a prepared lab. When biorisks are being managed properly, workers, the community, and the environment are safe and pathogens are secure from accidental or intentional release. Establishing observable measurements for the concepts of safety and security requires careful thought as well as clear documentation of the planning and preparation that goes into a successful biorisk management system.

Step 2: Define Outcome Indicators and Metrics

Outcome indicators are best determined before activity indicators. An outcome indicator is a demonstration that the outcome has occurred. If the desired outcome is that all personnel are competent in a certain procedure, one indicator may be that all personnel have completed training on how to properly perform the procedure. The accompanying metric would be the number of workers trained compared to the number untrained. Another desired outcome for biorisk management may be that a biorisk management committee has been established. An outcome indicator for this goal may be that the biorisk management committee has held meetings, and the metric can be the number of meetings held in a specified time period. Further indicators and metrics are required, of course, to measure if the committee is meeting the roles and responsibilities assigned.

Step 3: Define Activity Indicators and Metrics

Once outcome indicators are established based on goals and desired outcomes, activity indicators measure the existence of activities that have been put into place to assure the outcome. For example, to match the outcome indicator that workers are trained in a given procedure, the activity indicator could be that a training program for that procedure is established and available to all personnel. The metric may be a simple "Yes—the training program is available to all personnel" or "No—the training program is not available to all personnel." In the case of the biorisk management committee, an activity indicator might be that potential committee members have been invited to participate. The metric may be the affirmation (yes or no) that invitations have been sent to identified candidates. The value of pairing activity indicators with outcome indicators is that activity indicators are more often the responsibility of management rather than the workforce; thus, activity indicators hold leadership equally accountable for biorisk management as those working directly with pathogens or toxins. This parity of direct responsibility across all roles is a distinguishing characteristic of a biorisk management system.

Step 4: Collect Data and Report Indicator Results

As mentioned above, identifying the mechanisms for data collection and reporting should be pursued concurrently with indicator selection (outcome or activity indicator). However, merely deciding what indicators and metrics are needed does not assure that data is collected and reported in a manner that allows performance to be evaluated. Primary consideration in data collection involves determining who the data will be collected by, where it is to be collected, and when it is to be collected. Similarly, reporting procedures consider to whom the data will be reported, when it will be reported, and the mechanism by which the report is made. Care must be taken to assure that the collected and reported data is relevant and useful. Those to whom data is reported must have a clear understanding and use for that information relative to overall biorisk management system performance. For example, a worker may routinely log a reading from a piece of equipment to determine that work may safely and securely proceed. This same reading, in aggregate over time, may be used by program managers to assess equipment performance over time. The mechanism to allow the daily reading to be reported and to whom must be identified and in place so that it may be evaluated at the program level.

Some indicators require comparison to thresholds or tolerances, established by industry or organization best practices. A process and responsibility for obtaining these thresholds must be identified as part of the process for collecting data and reporting indicator results.

STEP 5: ACT ON FINDINGS FROM PERFORMANCE INDICATORS

Evaluating the results of performance indicators will require a response or action. Examples of the determinations to be made include:

- Which results meet expectations, exceed expectations, or fail expectations (or similar categorization), based on preestablished goals and objectives?
- Are results relevant to the preestablished goals and objectives?
- Are less than ideal results due to an unrealistic indicator or metric?
- Do results demonstrate an unanticipated but relevant result?
- Does any result require immediate corrective action (stop work, emergency procedures, retraining, etc.)?
- Does any result require preventive action (changes in assessment or mitigation strategies)?

Allowing a process to move forward as designed is a valid response if the performance indicators show that the desired outcome has been met or that satisfactory progress toward the desired outcome has been demonstrated. However, when the results from performance indicators show deviations from expectations, more significant action may be required. This is where the paired outcome and activity indicators are most helpful. For example, perhaps the outcome indicator shows that an unacceptably low number of personnel have been trained in a specific procedure. This could be due to personnel not attending a training session, or the accompanying activity indicator may show that a process to give all personnel access to the training has not been established or has deteriorated. This finding suggests that managers may need to evaluate and implement different training options or to increase training accessibility, or both. An evaluation of outcome-activity indicator pairings helps to determine the cause of the problem, but also helps to determine an appropriate response.

STEP 6: EVALUATE AND REFINE PERFORMANCE INDICATORS

Any system must include a periodic review of work procedures, processes, and other system functions to maintain reliability. In the case of performance indicators, several aspects must be reviewed by asking the following questions:

- Are current priorities still valid?
- Are the established indicators still needed (e.g., indicators may have shown that a goal has been reached, is satisfactorily being addressed, or even that the original priority may not be as important as previously thought)?
- Have established indicators provided the measurements necessary to demonstrate reliability?
- Do indicators provide the precision necessary to detect the performance desired? For example, does compliance in training translate to actual proficiency by users?
- Should additional indicators be added for new priorities or for previously identified priorities?
- Does new or emerging information or best practices shared from others need to be considered for incorporation in order to pursue continuous improvement of the system?

Once these questions are answered, the currently used performance indicators may require refinement. The same step-by-step process outlined above can be used to establish new, more appropriate indicators or to refine existing measurements.

SPECIFIC BIORISK MANAGEMENT EXAMPLES

The six steps presented above provide a framework for initiating and maintaining performance measurements as part of a biorisk management system. This section presents biorisk management examples based on the three key actions for implementing effective biorisk management performance indicators presented previously. These actions are repeated here:

1. Establish performance indicators during the planning stage—concurrently with the process of setting goals and objectives and assigning roles and responsibilities for each new project or program.
2. Involve persons at every level and function of an organization to conduct biorisk management performance measurements.
3. Evaluate and use existing data to create and expand performance measurements.

This section serves only as an example and is not fully inclusive of the various options available for use in performance indicators and metrics within a biorisk management system. The indicators and metrics listed in Tables 8.2 through 8.5 may or may not be relevant to every biorisk management system. Readers are encouraged to use the six steps outlined above and the examples given below as a starting point to develop their own indicators and metrics relevant to their facility. There is no shortcut or checklist for this work—going through this process will assure that performance measurements are more relevant to the program, which in turn will strengthen the biorisk management system.

For the first two examples, a biorisk management goal was chosen, using the provisions of the CWA 15793:2011 (European Committee for Standardization 2011) as a starting point. Objectives were designed to take steps toward meeting the goal, and in each example, performance indicators and metrics were developed to measure whether the stated objectives had been addressed and if, as a result, the overarching goal was closer to being achieved.

In Example A, the steps for including performance indicators as part of the planning process are shown. Example B builds on Example A, showing the possible expansion of performance measurements across an organization—at the lab, program, and organization levels.

Example C takes a common existing data collection procedure—the reading of a magnahelic gauge—and demonstrates how this single measurement might be used for different performance indicators across the organization, thus meeting different goals and objectives at different levels.

The last example, Example D, takes performance measurement out of the lab and into the field, listing some possible performance indicators for various field investigations. This example is included to demonstrate, along with other examples in this book, that biorisk management and the AMP model, specifically performance in this example, are relevant and applicable beyond the laboratory setting.

Example A: Setting Biorisk Management Performance Indicators during the Planning Stage

Example A, as outlined in Table 8.2, focuses on the critical function of a biorisk management system that includes the physical security of biological materials. One mechanism to achieve security is to assure that only authorized workers access these materials. A risk assessment should identify the materials and facilities for which strict authorized access is most critical. Mitigation strategies for implementing access controls include individually assigned keys, keycards, proximity or magnetic cards, or even biometric sensors (iris scan, hand geometry, fingerprint, etc.).

In the planning phase, where these risk assessments are completed and risk-based mitigation strategies are identified and proposed, performance indicators must also be established to provide documented assurance that the assessment is accurate and the mitigation is appropriate and effective. Figure 8.2 shows the basic flow of the plan-do-check-act aspects of a biorisk management system. Note that performance indicators are established during the planning phase and that they are based on the goals, objectives, roles, and responsibilities of a biorisk management system. The CWA 15793:2011 document (European Committee for Standardization 2011) is a

TABLE 8.2

Example A: Developing Performance Indicators from an Identified Biorisk Management Goal and Objective

Step 1: Identify Key Issues of Concern

BRM function: Operational controls.

Subfunction: Security.

Goal: Controls are established, based on biorisk assessment, for the physical security of cultures, specimens, samples, and potentially contaminated materials or waste (CWA 15793 4.4.4.8.4).

Objective: Assure that facilities and equipment (where valuable biological materials (VBMs) are used, stored, or disposed) are accessed only by authorized (screened, trained, and assigned) workers.

Purpose of indicators: To determine if unauthorized persons are gaining access or attempting to gain access to VBMs, potentially increasing the risk of theft, misuse, or accidental release.

Steps 2 and 3: Define Outcome and Activity Indicators and Metrics		Step 4a: Collect Data			Step 4b: Report Indicator Results		
Outcome	Activity	Who Measures?	Where Is It Measured?	When Is It Measured?	To Whom Are Results Reported?	When Are Results Reported?	How Are Results Reported?
Indicator: How many unauthorized attempts to access facilities or equipment where VBMs are used, stored, or disposed were reported? Metrics: (a) Number of reports from workers of unauthorized persons stopped, by workers, before entry or use; (b) number of reports from access control logs of attempted entry with unauthorized credentials.	Indicator: Do lab (or organization) SOPs require reporting attempts by unauthorized users to enter or use facilities or equipment where VBMs are used, stored, or disposed? Metric: Yes or no; if yes, include copies of SOPs. Indicator: Is someone assigned to review log from access control equipment (e.g., card readers, etc.) to determine if unauthorized access was attempted? Metric: Yes or no; if yes, list name(s) of assigned personnel.	Lab personnel—as assigned by laboratory director	Specific facilities and equipment where VBMs are used, stored, or disposed	Ongoing	Laboratory director	Outcome: At the time any lab personnel report attempted unauthorized access and upon weekly review of access control logs. Activity: Annually, when lab SOPs are reviewed.	Verbal notification of attempted unauthorized access, followed by written report. Written report on SOP review. Documented assignment of access control log review.

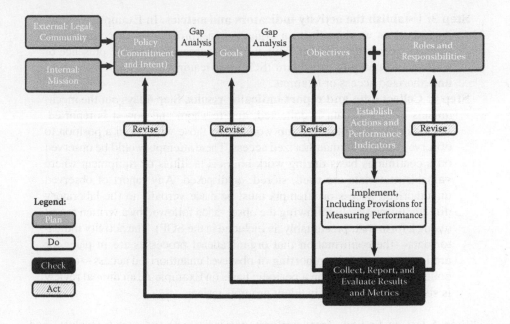

FIGURE 8.2 Plan-do-check-act flow in biorisk management performance.

good starting point for biorisk management system goals. In addition, CWA 16393: Laboratory Biorisk Management—Guidelines for the Implementation of CWA 15793:2008 (European Committee for Standardization 2012) provides further guidance on the implementation of a biorisk management system.

For Example A, the stated goal is based on CWA provision 4.4.4.8.4 and is paraphrased as "controls are established, based on biorisk assessment, for the physical security of cultures, specimens, samples, and potentially contaminated materials or waste" (European Committee for Standardization 2011). The objective, based on the notes section of the same provision, "assure that facilities and equipment (where valuable biological materials (VBMs) are used, stored, or disposed) are accessed only by authorized (screened, trained, and assigned) workers," is one step that can be taken to meet the goal. Both the goal and objective(s) can be listed as part of:

Step 1: Identify key issues and concerns. A further refinement is to state the purpose of the indicators. In this example, the purpose is "to determine if unauthorized persons are gaining access or attempting to gain access to VBMs, potentially increasing the risk of theft, misuse, or accidental release."

Step 2: Define the outcome indicators and metrics. A potential desired outcome for achieving the goal and objective would be the detection of unauthorized access or attempts. That outcome could be measured by reports of observed unauthorized access or access attempts, or recorded by electronic logging devices that monitor access or access attempts.

Step 3: Establish the activity indicators and metrics. In Example A, activity indicators would reflect an organizational process for detecting unauthorized access or attempts. This might be represented by the presence of a standard operating procedure (SOP) that requires reporting of observed unauthorized access or attempts.

Step 4: Collect data and report indicator results. Step 4 lays out the mechanisms by which data is collected, and to whom and how it is reported. For Example A, authorized lab workers are those who are in a position to observe attempts at unauthorized access. These attempts would be observed on a continuous basis during work hours at facilities or equipment where valuable materials are used, stored, or disposed. Any report of observed unauthorized access or attempts must be made verbally to the laboratory director immediately following the observation followed by a written report (using a form that, presumably, is included in the SOP). The activity indicator data—the confirmation that organizational processes are in place (or are lacking) to require reporting of observed unauthorized access—can be collected and reported on a periodic basis (in Example A, an annual review is suggested), unless an incident prompts review.

Step 5 (act on findings from performance indicators) and step 6 (evaluate and refine performance indicators) cannot be presented here without reported data, and thus are not listed in the table. However, once data is collected from outcome and activity indicators, the person who receives the reports (in Example A, this is the laboratory director—this will vary at actual institutions) must review the reports to determine if results match the desired outcomes and if defined activities are in place. Results not matching expectations may require corrective or preventive action. No change may be necessary for results that indicate satisfactory progress. As with mitigation strategies, no single performance indicator is adequate for monitoring performance. Multiple indicators increase the coverage of the different factors that influence the success, or lack of success, of mitigation strategies.

If results indicate that the system for assuring unauthorized access is functioning as desired, decreases in the frequency of measurement or reporting might be acceptable or the measurements can be left in place. Alternately, additional indicators or more stringent outcomes should be put in place to further mitigate identified risks. If the measurements do not appear to be the correct measurements to detect unauthorized access, the indicators should be changed or refined to better indicate performance.

EXAMPLE B: EXPANDING BIORISK MANAGEMENT PERFORMANCE INDICATORS DURING THE PLANNING STAGE

Example B is shown in Table 8.3. Example B uses the same biorisk management scenario as Example A—the physical security of biological materials with the same goal and objective, assuring only authorized access—but because biorisk management responsibility exists in every level and role of an organization, Example B

TABLE 8.3

Example B: Developing Performance Indicators for Different Levels of an Organization

Step 1: Identify Key Issues of Concern

BRM function: Operational controls.

Subfunction: Security.

Goal: Controls are established, based on biorisk assessment, for the physical security of cultures, specimens, samples, and potentially contaminated materials or waste (CWA 15793 4.4.4.8.4).

Objective: Assure that facilities and equipment (where valuable biological materials (VBMs) are used, stored, or disposed) are accessed only by authorized (screened, trained, and assigned) workers.

Purpose of **lab-level** indicators: To determine if unauthorized persons are gaining access or attempting to gain access to VBMs, potentially increasing the risk of theft, misuse, or accidental release.

Steps 2 and 3: Define Outcome and Activity Indicators and Metrics		Step 4a: Collect Data			Step 4b: Report Indicator Results		
Outcome	Activity	Who Measures?	Where Is It Measured?	When Is It Measured?	To Whom Are Results Reported?	When Are Results Reported?	How Are Results Reported?
Indicator: How many unauthorized attempts to access facilities or equipment where VBMs are used, stored, or disposed were reported?	Indicator: Do lab (or organization) SOPs require reporting attempts by unauthorized users to enter or use facilities or equipment where VBMs are used, stored, or disposed?	Lab personnel—as assigned by laboratory director	Specific facilities and equipment where VBMs are used, stored, or disposed	Ongoing	Laboratory director	Outcome: At the time any lab personnel report attempted unauthorized access and upon weekly review of access control logs.	Verbal notification of attempted unauthorized access, followed by written report.
Metrics: (a) Number of reports from workers of unauthorized persons stopped, by workers, before entry or use; (b) number of reports from access control logs of attempted entry with unauthorized credentials.	Metric: Yes or no; if yes, include copies of SOPs.						Written report on SOP review.
	Indicator: Is someone assigned to review log from access control equipment (e.g., card readers, etc.) to determine if unauthorized access was attempted?					Activity: Annually, when lab SOPs are reviewed.	Documented assignment of access control log review.
	Metric: Yes or no; if yes, list name(s) of assigned personnel.						

continued

TABLE 8.3 (Continued)
Example B: Developing Performance Indicators for Different Levels of an Organization

Purpose of **program-level** indicators: To determine which labs, based on biorisk assessment, require more stringent access controls and to assure that facilities and equipment that carry a higher biorisk are protected at a more stringent level than those with lower risk.

Outcome	Activity	Who Measures?	Where Is It Measured?	When Is It Measured?	To Whom Are Results Reported?	When Are Results Reported?	How Are Results Reported?
Indicator: What is the percentage of labs where biorisk assessments are up to date? Metric: Number of facilities, compared to total number of facilities, where biorisk assessments have been completed in the past year. Indicator: How many labs have been determined, based on the current biorisk assessment, to require access controls? Metric: List of labs, their relative risk level, and the level of access control required (assigned and installed).	Indicator: Does a written organizational process and requirement for conducting biorisk assessments exist? Metric: Yes or no; copy of process. Indicator: Has the organization developed and implemented a process by which access control measures are assigned and installed based on biorisk level? Metric: Yes or no; copy of process. Indicator: Have the processes described above been reviewed and revised, as necessary, in the previous 2 years? Metric: Yes or no; dates of most recent review/revision.	Program level	Across the organization	Routinely and when new facilities, equipment, procedures, or VBMs are introduced	Top management (or biorisk management oversight body)	Quarterly	Written summary

continued

TABLE 8.3 (Continued)
Example B: Developing Performance Indicators for Different Levels of an Organization

Purpose of organization-level indicators: To determine which workers require which level of access to facilities, equipment, and VBMs. To assure that workers who work or may need to work in facilities where access control is required are evaluated to determine the appropriate level of access, and VBMs required to conduct their work. To assure that workers (and others) with limited or no need to access are excluded or are provided access in a more restricted manner.

Outcome	Activity	Who Measures?	Where Is It Measured?	When Is It Measured?	To Whom Are Results Reported?	When Are Results Reported?	How Are Results Reported?
Indicator: How many workers, relative to entire workforce, have been evaluated to determine their level of access? Metric: Number of workers evaluated; number of workers in the entire workforce. Indicator: How many workers have been placed in each access control zone? Metric: Number of workers in each access control zone.	Indicator: Does an organizational process exist that requires that workers who work or may need to work in facilities where access control is required are evaluated to determine the appropriate level of access? Metric: Yes or no; copy of process. Indicator: Is someone assigned to make access determinations? Metric: Yes or no; name(s) of assigned personnel.	Organization level	Across the workforce	Hiring, reassignment (at lab or organization level), termination	Top management	Annually, or less frequently	Written summary

expands performance measurements across the broader organization. The first indicator purpose—targeted at the lab level—is the same as in Example A and is therefore duplicated from Table 8.2. Two additional purposes are defined to meet the same objective, but these are targeted toward program (often managed by a biorisk management advisor) and organization (managed by organization directors) levels. The same four steps as defined in Example A are used; however, because of the different level and roles, different outcomes and activities are identified.

For indicators typically measured by and assigned at the program level (e.g., biorisk management advisor, biorisk management advisory committee, security office, safety office), the purpose of the indicator is the determination of which labs, based on the biorisk assessment, require more stringent access controls. The indicators focus, at this level, on biorisk assessments. Have they been performed? At which labs? Does an organizational process and requirement for conducting biorisk assessments exist? Which labs, based on the biorisk assessment, require more stringent access controls? Is there an organizational process to assign and install appropriate access controls based on the biorisk assessment?

For the organizational level, the focus shifts to personnel. Personnel issues are generally handled through offices that report to top management (e.g., human resources), and biorisk management indicators that focus on personnel are thus appropriately assigned to those top managers. At this level, indicators focus on whether personnel are evaluated to determine their suitability for access (screened and trained), and if they are appropriately assigned to "matching" access levels.

All three levels contribute significantly to meeting the stated objective (under step 1) of assuring that valuable biological materials are accessed only by authorized persons. This objective, in turn, contributes to meeting the goal of establishing controls for the physical security of those materials. Obviously, there are many other goals, objectives, and performance indicators that can be used in combination (or alternately) with these first two examples, but these provide a snapshot of what a performance measurement initiative might look like.

Example C: Utilizing Existing Data to Create or Expand Performance Measurements

Example C is shown in Table 8.4. This example focuses on utilizing data that is already being collected for other purposes and applying it to provide performance information across the whole organization. In this example, the existing information is the reading on the magnahelic gauge found on most biological safety cabinets (BSCs). This gauge measures the pressure difference across the high-efficiency particulate air (HEPA) filters between the work space and the inner plenum of the cabinet. Changes in readings over time can indicate a problem with the HEPA filter, which can result in improper function of the BSC. Most lab workers are trained to check the magnahelic gauge before beginning work to assure that the filter and BSC are working within specifications. The data from this simple reading, however, can be used for many other performance measurements within the institution depending on who uses the data. Table 8.4 provides examples of using this data for

TABLE 8.4

Example C: Using the Same Data Source for Different Performance Indicators and Metrics across Multiple Levels of an Organization

BRM function: Operational controls.
Subfunction: Equipment.

Who	What	Where	When	Why	Example Indicators	
					Outcome	How / Activity
Lab level	Magnahelic gauge pressure reading on BSC	Individual BSC in laboratory	Daily	Is BSC functioning within specifications so that work can proceed?	Indicator: Is magnahelic reading within predefined acceptable working levels? Metric: Actual reading and position relative to predefined acceptable working levels.	Indicator: Do lab SOPs require reading of magnahelic gauge before work begins? Metric: Yes or no. Indicator: Are workers trained how to read magnahelic gauge? Metric: Yes or no.
Program level	Magnahelic gauge pressure reading on BSC	BSCs within lab and organization	Periodic	Are there HEPA filters in the lab's or organization's BSCs that are nearing time for replacement?	Indicator: Is magnahelic reading nearing predefined range that indicates end of life? Metric: Actual reading and position relative to benchmarked end-of-life reading.	Indicator: Has an end-of-life range been researched and documented (within organization or from industry specifications) to benchmark HEPA end-of-life readings? Metric: Yes or no; documented process and results.
Organization level	Magnahelic gauge pressure reading on BSC	BSCs within organization	Annual	How many HEPA filter changes should be budgeted in the next fiscal cycle?	Indicator: How many magnahelic readings across the organization, combined with benchmark data, indicate that the HEPA filter may need to be changed? Metric: Number of readings indicating possible replacement.	Indicator: Is the organization routinely and proactively budgeting for replacement of HEPA filters in BSCs? Metric: Yes or no; annual budget over time. Indicator: Are resources and responsibilities allocated so that data is collected in a manner to detect trends in expected end-of-life status for HEPA filters in BSCs? Metric: Yes or no; list resources and responsibilities.

programmatic and organizational performance assessments. The program level can use aggregated data from individual readings to determine whether HEPA filters are nearing their end of life. This requires that the data associated with each device be recorded and reported, in some fashion, to the program level for aggregation and review. Evaluating the program-level outcome indicators also requires that some process for determining end-of-life thresholds is in place; this is one type of activity indicator that could be established and measured. The data aggregated and evaluated by the program level can, in turn, be reported to the organizational level to assure that budgets are in place for expected HEPA filter replacement.

This example is just one of many where existing data can be used to improve biorisk management system performance. Evaluating and employing existing data sources for their utility beyond their current use may allow an organization to see larger-picture results.

EXAMPLE D: USING BIORISK MANAGEMENT FUNCTIONS AND PERFORMANCE INDICATORS OUTSIDE OF LABORATORIES

The concept of biorisk management is most often associated with the activities of laboratories where biological agents and toxins are used, stored, or disposed. However, there are many other settings where biological agents and toxins may be encountered and the pillars of biorisk management—assessment, mitigation, and performance—are equally applicable in these settings.

Example D is shown in Table 8.5. This example focuses on an activity common to the field collection of samples: safe and secure transport. Samples collected outside of laboratories from farm animals or wildlife or from rural human populations are often collected in settings with little containment infrastructure and then are transported from the sample site to a laboratory for analysis. During transport, the samples may be transferred to several different carriers. In this scenario, assuring safe and secure transport focuses on also assuring that the documentation that accompanies the sample is accurate and authentic, and that those who participate in transport are authorized to do so. Table 8.5 is formatted identically to the previous tables using the steps for establishing and implementing performance indicators. The purpose of the example indicators is to verify sample identity during field collection and transport. The outcome indicators measure the validity of documentation that accompanies the samples, as well as visual affirmation that the sample matches expected appearances (blood, tissue, swab, etc.). The activity indicators measure that processes are in place to require documentation and to verify authorized transport and carriers. Note that the outcome indicators, for this example, are measured at the point of transfer, not at the laboratory (unless the lab is at a point of transfer). In another situation, the outcome could be the condition of the package and sample in the receiving area.

CONCLUSION

Measuring biorisk management system performance is critical to assuring that pathogens and toxins are handled safely and securely in any setting. Currently, biorisk

TABLE 8.5
Example D: Developing Performance Indicators from an Identified Biorisk Management, but Nonlaboratory Goal and Objective

Step 1: Identify Key Issues of Concern

BRM function: Operational controls.

Subfunction: Transport of biological agents and toxins.

Goal: Procedures for the safe and secure transport of culture, specimens, etc., are established and maintained (CWA 15793 4.4.9).

Objective: Assure chain of custody for samples that documents traceability of material movements, authorized movement of materials, and movement by trustworthy carrier(s).

Purpose of indicators: To determine if sample identity can be verified at different time points from collection to transport to receipt.

Steps 2 and 3: Define Outcome and Activity Indicators and Metrics

		Step 4a: Collect Data				Step 4b: Report Indicator Results	
Outcome	Activity	Who Measures?	Where Is It Measured?	When Is It Measured?	To Whom Are Results Reported?	When Are Results Reported?	How Are Results Reported?
Indicator: Is the sample accompanied by unaltered and complete documentation that indicates the source of the sample, the date and time of collection, and the person(s) to whom custody of the sample has been given or transferred? Metrics: Yes or no. Indicator: Is the sample(s) accompanied by unaltered and complete documentation that indicates the source of the sample, the date and time of collection, and the person(s) to whom custody of the sample has been given or transferred? Metrics: Yes or no.	Indicator: Does a process exist that requires documentation of key information regarding the sample and the identity of those to whom custody of the sample is entrusted? Metric: Yes or no; SOP. Indicator: Does a process exist to easily and quickly verify that sample movement and the carrier are authorized? Metric: Yes or no; SOP. Indicator: Does a process exist for reporting suspicious or incomplete documentation? Metric: Yes or no.	Outcome: recipient. Activity: Process owner as designated by management (often lab manager or biorisk management advisor).	Outcome: At point of transfer, before assuming custody. Activity: At location of periodic review of documentation.	Outcome: At time of transfer, before assuming custody. Activity: At least annually, or following suspicious activity.	Outcome: The recipient receives the report from the previous holder and also documents that the transfer is made according to procedure. If suspicious or incomplete, the recipient reports immediately to designated supervisor. Activity: No less frequent than annually, or following suspicious activity.	Outcome: At time of transfer. Activity: No less frequent than annually, or following suspicious activity.	Signature of each recipient indicates that they have accepted documentation as unaltered and complete. Immediate verbal report, followed by written report, to designated supervisor to indicate concerns. Activity: By process owner to designated manager or oversight body (e.g., biorisk management advisory committee).

performance measurements are rarely executed, or they are used only after an incident occurs. Relying on failure data is inappropriate, ineffective, and may significantly impact the lives and welfare of employees and others. The implementation of thoughtful performance indicators requires careful planning, but the measurements themselves need not be difficult to gather or costly. Performance measurements must be established during planning concurrently with goals, objectives, roles, and responsibilities. The six steps to establishing effective performance indicators during this planning period are: (1) identify the key issues of concern, (2) define outcome indicators and metrics, (3) define activity indicators and metrics, (4) collect data and report indicator results, (5) act on findings from performance indicators, and (6) evaluate and refine performance indicators.

All levels and roles within an organization are responsible for biorisk management, including measuring performance. Different roles and responsibilities require different types of indicators to assure biorisk management performance. The use of both outcome and activity indicators helps to identify both top-down and bottom-up successes or weaknesses. One of the first steps in establishing a biorisk management performance initiative is to evaluate and exploit, if appropriate, existing data sources. Rather than being overwhelmed by setting performance measurements for all aspects of biorisk management, managers should focus on a few key areas first, and then gradually add more areas. Success in both establishing indicators and metrics and also in measuring positive performance will breed further success and positive performance. This positive performance will further reinforce risk reduction strategies and support a safe and secure workplace.

REFERENCES

Choudry, R.M., D. Fang, and S. Mohamed. 2007. Developing a Model of Construction Safety Culture. *Journal of Management in Engineering*, 23(4): 207–212.

DeJoy, D.M., R.R.M. Gershon, and B.S. Schaffer. 2004. Safety Climate: Assessing Management and Organizational Influences on Safety. *Professional Safety*, 2004: 50–57.

DeVries, J.E., M.M. Burnette, and W.K. Redmon. 1991. AIDS Prevention: Improving Nurses' Compliance with Glove Wearing through Performance Feedback. *Journal of Applied Behavior Analysis*, 24(4): 705–711.

Eherts, D.M. 2008. Lessons Learned from Aviation Safety. *Journal of Safety Research*, 39: 141–142.

European Committee for Standardization. 2011. CEN Workshop Agreement (CWA) 15793: Laboratory Biorisk Management Standard.

European Committee for Standardization. 2012. CEN Workshop Agreement (CWA) 16393: Laboratory Biorisk Management—Guidelines for the Implementation of CWA 15793:2008.

Fernandez-Muniz, B., J.M. Montes-Peon, and C.J. Vazquez-Ordas. 2007. Safety Culture: Analysis of the Causal Relationships between Its Key Dimensions. *Journal of Safety Research*, 38: 627–641.

Gershon, R.R., C.D. Karkashian, J.W. Grosch, L.R. Murphy, A. Escamilla-Cejudo, P.A. Flanagan, E. Bernacki, C. Kasting, and L. Martin. 2000. Hospital Safety Climate and Its Relationship with Safety Work Practices and Workplace Exposure Incidents. *American Journal of Infection Control*, 28(3): 211–221.

Glendon, A.I., and D.K. Litherland. 2001. Safety Climate Factors, Groups Differences and Safety Behaviour in Road Construction. *Safety Science*, 39: 157–188.

Harper, D. 2014. Performance. Dictionary.com. Online Etymology Dictionary. http://dictionary.reference.com/browse/performance (accessed June 20, 2014).

Jovasevic-Stojanovic, M., and B. Stojanovic. 2009. Performance Indicators for Monitoring Safety Management Systems in Chemical Industry. *Chemical Industry and Chemical Engineering Quarterly*, 15(1): 5–8.

Kirkpatrick, D.L., and Kirkpatrick, J.D. 1994. *Evaluating Training Programs*. San Francisco, CA: Berrett-Koehler Publishers.

Neal, A., and M.A. Griffin. 2006. A Study of the Lagged Relationships among Safety Climate, Safety Motivation, Safety Behavior, and Accidents at the Individual and Group Levels. *Journal of Applied Psychology*, 91(4): 946–953.

OECD Environment Directorate. 2008. *Guidance on Developing Safety Performance Indicators Related to Chemical Accident Prevention, Preparedness and Response*. Vol. 19. Paris: OECD Environment, Health, and Safety Publications.

Reniers, G.L.L., B.J.M. Ale, W. Dullaert, and K. Soudan. 2009. Designing Continuous Safety Improvement within Chemical Industrial Areas. *Safety Science*, 47: 578–590.

Turnberg, W., and W. Daniell. 2008. Evaluation of a Healthcare Safety Climate Measurement Tool. *Journal of Safety Research*, 39: 563–568.

Wikipedia. 2013. Management System. December 12. http://en.wikipedia.org/w/index.php?title=Management_system&oldid=585777611 (accessed June 20, 2014).

Wu, T.-C., C.W. Liu, and M.C. Lu. 2007. Safety Climate in University and College Laboratories: Impact of Organization and Individual Factors. *Journal of Safety Research*, 38: 91–102.

9 Communication for Biorisk Management

Monear Makvandi and Mika Shigematsu

ABSTRACT

This chapter discusses hazard, risk, and crisis communication within the context of biorisk management. Several main aspects exist when delivering hazard, risk, and crisis communication: the type of information communicated, the urgency of the communication, and the media used for the communication. Communication goals should be identified based on the audience, the content of the message, and the nature of the hazard, risk, or crisis. There may be multiple audiences with differing information needs for the different types of message (hazard, risk, or crisis). The messages should be created to address the different audiences' needs and how well the audience understands the hazard, risk, or crisis that needs to be communicated. Many audiences will internalize the information subjectively, and if their particular concerns are not addressed in the message, they may presume a worst-case scenario even when presented with complete information. Failure by institutions to incorporate value judgments into communication about hazards, risks, and crises can distract from and skew the message's original intent—which could range from threshold hazardous biological substance reporting to a full-scale alert in the event of a dangerous release. Including the public as a target audience and a source of biorisk management information, such as information on threats for assessment purposes or contamination for performance measures, is often overlooked during the risk assessment process. Biorisk management systems should incorporate a structured communication process to discuss and evaluate hazards, risks, and crises to mitigate consequences of negative public perception of the risk that the facility and the hazards contained therein pose to the community.

A CASE STUDY

In the aftermath of the September 11, 2001, terrorist attack against the United States, the anthrax letter mailings (Amerithrax) that targeted US media and government workers, killing 5 people and sickening 17, raised the worldwide fear of biological terrorism. The United States and other countries had renewed concerns about the misuse of biological agents, expertise, and materials and brought these issues to the forefront of national and international security organizations. One outcome of the investigations into Amerithrax was the recognition that the world, and the United States in particular, lacked available and adequate research facilities to study emerging infectious diseases (National Institute of Allergy and Infectious Diseases

2004). The National Institute of Allergy and Infectious Diseases (NIAID) and a federal independent panel of experts developed a strategic plan for biodefense research aimed at developing new and improved diagnostics, vaccines, and treatments for diseases caused by infectious agents to address this research gap (National Institute of Allergy and Infectious Diseases 2007).

To address the national shortage of containment and high-containment laboratory space, the strategy included plans to construct and renovate laboratories across the country that were located in areas where cutting-edge scientific research and development was already occurring and where scientific research communities already existed and collaborated. The National Emerging Infectious Diseases Laboratories (NEIDL) was one of four new construction projects, and in 2003, the Boston University and Boston Medical Center site was selected in a nationwide selection process (National Institute of Allergy and Infectious Diseases 2010). However, what happened after site selection involved a series of missteps by the technical community in communicating the justification and need for the NEIDL facility, and misunderstanding the community perception of the planned work, the inherent safety and security, and the selection process.

Groups opposed to operating a high-containment laboratory in a congested urban area have focused on a lack of government transparency, work with deadly diseases that have no known cures, and structural violence against marginalized populations. Among the earliest arguments regarding social justice was the decision to site the laboratory in the South Boston neighborhood. The NIH expert panel evaluated the environmental justice based on three federal criteria and determined that it met two of these three criteria (National Institute of Allergy and Infectious Diseases, NIH 2005).

More recently, a second social justice argument was raised that emphasizes that the facility's mission does not address existing public health concerns within the South Boston neighborhood. Instead of researching diseases of local concern (perhaps HIV/AIDS and hepatitis), the new facility plans to conduct research on diseases of global concern, such as Ebola, Hendra, anthrax, severe acute respiratory syndrome (SARS), and novel viral agents, all diseases that are not endemic in the local population, and therefore do not immediately threaten the health security of the community. The community is concerned that the nonendemic diseases, if released, would introduce new health concerns capable of harming or possibly decimating the local population; social justice advocates further argue that the community will receive no benefit from the scientific advances generated by this type of research. Instead of addressing local diseases of concern, the argument faults the laboratory for its intended focus on the exotic diseases that are not endemic and do not immediately threaten the health security of the community. Opponents of the research refer to the lab disparagingly as the Boston bioterror lab; they evoke a menacing image of a laboratory shrouded in secrecy that would create offensive biological weapons and resurrect the long-abandoned practice of utilizing vulnerable local populations as test subjects without informed consent (Cedrone 2012). NEIDL has stated that first, bioweapons work is illegal, and second, there will be no classified work conducted at NEIDL (Boston University Medical Campus National Emerging Infectious Diseases Laboratory, n.d.).

In designing and evaluating the lab, including the work of several environmental impact and blue-ribbon panels that assessed the potential risk of the facility to the South Boston neighborhood, the negative public perception was not considered as a hazard that could affect the normal operations. Failure to recognize the public perception problem—not a lack of available scientific risk assessment data and the related safety and security mitigation measures—means the laboratory is still engaged in a public relations campaign that seeks to dispel the negative views that the new lab is unsafe. This campaign has included tours, demonstrations of safety and security features, creation of a public liaison community group, multiple risk assessments that focus on accidental release of an agent, and recommendations for mitigation measures that address the risk of release of an agent. In addition, the NEIDL has received certification by not only the NIH Environmental Safety Review, but also the Boston Public Health Commission and other agencies.

After a long legal battle, in January 2013, the Boston Health Commission approved work to start in the BSL3 labs. Subsequently, in September 2013, federal courts ruled that the Final Supplementary Risk Assessment prepared by experts on behalf of the National Institutes of Health (NIH) analyzed the risks associated with pathogen research in BSL3 and BSL4 laboratories adequately and that the research could be conducted safely at the NEIDL site. Massachusetts State Court also ruled that the Supplemental Final Environmental Impact Report complied with Massachusetts Environmental Policy Act and dismissed a suit (Boston University Medical Campus National Emerging Infectious Diseases Laboratory, n.d.).

Despite volumes of scientific evidence and a detailed, risk-based approach to safety and security, the NEIDL faces an uphill battle to occupy and operationalize the containment and high-containment research suites (National Academies 2010). There have been several lawsuits and proposed city ordinances to ban all high-containment research in Boston. When the scientific argument demonstrating adequate safety and security failed to convince community members opposed to the laboratory, proponents for opening NEIDL turned their messaging focus to highlight the beneficial economic impacts, including jobs and how growth of the laboratory could offer benefits, such as in the local community. Both of these public relations strategies failed to address the social justice concerns and perceived risk by those opposed to the lab's high-containment operations.

The disconnect between the technical evidence demonstrating safety and security and the stalwart position from those opposed lies in risk perception. The proponents must address the perceived risks in a manner that addresses the drivers of perception instead of repeating the refrain, "Trust us, we're the experts."

Laboratory facilities should consider the risk that negative community perception poses to the laboratory's ability to operate safely and securely in its baseline risk assessment (violent protests, sabotage, etc.), rather than only evaluating the risk the laboratory itself poses to the environment and community. In the case of NEIDL, a portion of the community views the laboratory as an unacceptable hazard. One mitigation strategy to address the risk of negative community perception on the capability of the facility to conduct safe and secure laboratory operations is a communication plan based explicitly on risk assessment that is incorporated into the facility biorisk management system.

BACKGROUND

Members of society require communication to build good human interactions. Risk communication was initially developed as a means to investigate how the best expert assessments could be communicated to the public in order to bridge the conflict between public perceptions and expert judgment on the risks (Golding et al. 1992). Risk communication serves to educate others about the risks, and can help change attitudes. Even though the initial intent was to educate the public, in reality, advocacy or announcement is often the last purpose of risk communication. Biorisk management systems need to consider the public as part of the system of risks to mitigate. No matter how carefully crafted, the one-way communication process of conveying a message to the public in persuasive language produces little effect (Leiss 2004). Good risk communication can help to alleviate fears regarding a risk, or inversely, can help to demonstrate that a risk is unacceptable. Risk communication should become part of a structured risk assessment process.

When communication on the risks is initiated, the process starts with the core message about risk that will explain both the added values and the levels of risk that come with the project as interpreted by personal experience and perspective. The intent is to make it understandable for the public, other stakeholders, or a specific target audience. Once the message has reached the target audience, the sender has an obligation to verify with the audience if the message received is the same as the original message that the sender intended. This complex process often fails when irrelevant information, or noise, interrupts the communication channels, or because it is misunderstood by the receiver. Although the term *communication* denotes the idea of sending out messages to a particular audience, the more important elements of the communication are on the receiving end. The ability to receive and process the messages can be used as performance measures for the system's ability to mitigate the perceived risks.

Understanding the audience is important in biorisk management systems, especially when communicating risks. One can know the audience and stakeholders by listening to their views, opinions, and problems, and collecting information from them. It is critical to learn where they see value prior to sending out messages about requirements or changes. Listening to the audience provides a baseline understanding of what the perception of the hazard or risk is. It also enables the message to be understood in a way that avoids misinterpretation that could result from using words that have different meanings among the groups involved. It is also necessary to choose the right communication method to reduce the noise—distracting from the message's point. In most cases, this means it is better to have direct communication with the intended audience (such as holding a stakeholder meeting with a question and answer session), but sometimes it is better to judiciously construct and deliver a written message (Beecher et al. 2005). If the message is not heard, or it is ignored, the actual risk is not communicated regardless of how transparent and timely it is. The choice to hear the message is in the hands of the audience, but the risk communicator has the responsibility and need to communicate, including verification the message was received as intended.

The NEIDL risk communication strategy has failed to address the public perception gap; this gap is the space between the audience's risk perception and exaggerated fear of a hazard, and the rational, scientific facts of that hazard. The public perceived the risk of conducting biodefense research in terms of the hazards and agents involved in the research, the history of past research for offensive biological weapons, and the popular information available about the diseases to be studied. Communication efforts have not focused on the difficulty of communicating information on risk. When the public learns about the probability of a release of agents associated with the laboratory, the public immediately wants to know what would happen if the low-likelihood event occurred, not what mitigation measures have been implemented to reduce the likelihood further (Slovic 1980). Later in this chapter, we address lack of planning for catastrophic failure and crisis communication. The issue is raised here to illustrate that the information communicated may not address the audience's perceived risk or concerns.

ROLES AND RESPONSIBILITIES

In general, all personnel in the facility should have a good factual understanding of the actual risks posed from the laboratory's operations to both workers and the community; however, specific roles and responsibilities must be defined relative to the communication of hazard, risk, and crisis information. The following describes typical facility positions and community roles:

Biorisk management advisors: The individual in charge of biorisk management relies heavily of frontline staff, management, and leadership to provide accurate technical feedback on identified hazards and the subsequent risks. These individuals act as the liaison and communicator to link hands-on, frontline laboratory workers and contractors to laboratory managers, directors, executive management, and other stakeholders. These people are often the primary risk assessors, who perform the role of consultants for management decisions on implementation of mitigation measures. This role is expected to have the most in-depth understanding of the results of the risk assessments.

Principal investigators/scientists/researchers: These individuals provide the information and data that form the basis of the risk assessments. They are expected to understand the rationale, policy, and priority of the assessment, as well as the importance and practical execution of the mitigation measures. Their understanding and support of the risk assessment is crucial for effective biorisk management.

Security personnel: These individuals are security experts who may provide valuable insight into risk assessments. They may be involved in the implementation of biosecurity mitigation measures (e.g., providing security for key areas), or they may act as an inspector to check their functionality and performance. The need to prioritize a potential adversary's means, motive, and opportunity to access the facility for the purpose of theft or sabotage is a particularly important role for the security professionals.

Legal consultant or department/labor safety officer: These individuals usually have no direct involvement with the risk assessment process, but their expert opinion is valuable when the message needs to be distributed to workers and the surrounding community to enhance their understanding and support. During the process of prioritizing risk or mitigation measures, their opinions may need to be considered. These individuals generally have unique backgrounds, and are often not intimately familiar with laboratory functions. Thus, the risk assessment team will need to ensure good communication and understanding with these individuals.

Contractors and custodial staff: These individuals are in the directly affected population with possibly limited knowledge about the hazards to which they may be exposed. Communication needs to start with having conversations to understand their level of knowledge, hear their concerns, and establish trust to obtain their support for implementation.

Executive management: These individuals should be the decision makers and end users of the risk assessment results. They must have a substantive understanding of the results of the risk assessments so that they can make well-informed mitigation decisions. Resource allocation and financial support from this group are vital to establish biosafety and biosecurity mitigation measures. Often miscommunication of the results happens because this group may have inadequate understanding of the hazards and risks. Dialogue with executive management should be initiated in the early stage of the risk assessment, and by engaging and involving this group in the assessment itself. Translating the risk assessment results into general terms and practical steps will support executive management's understanding and decision making. This group is ultimately responsible for public communication in conjunction with public information officers. For example, executives and institutional leaders often make public statements when events gather media attention.

Administration: These individuals have limited access to laboratory areas, but daily access to the people who work in the laboratory. In general, risk communications directed to individuals who are not directly involved with the lab's technical work should be written for an audience with the assumption that they have little or no biotechnical background; therefore, the text should omit unnecessary scientific details, and complex subjects should be translated into plain understandable language. Biosecurity measures may significantly affect administrative personnel, as these individuals will likely have primary responsibility for ensuring or overseeing the facility's compliance with site-specific rules and regulations. This group should also understand the social justice impact of the facility's operation on the community, and understand the differences in perception of the hazards and risks the facility introduces into the community.

Community stakeholders: In addition to the local community where the facility is located, the families of workers and visitors should also be considered in the stakeholder category. They need to be informed of the hazards and the inherent risks that relate to their level of interaction with the facility and its staff. Additionally, certain community stakeholders, such as law enforcement and fire and medical responders, should be alerted to probable exposure risks for early detection purposes and incident response planning. The messages provided to community stakeholders and response groups should also incorporate a feedback loop so the facility can accurately address questions and clarify perceived risks and concerns.

Public and media relations/public information officer: These individuals have responsibility for listening to community members, as well as understanding the general tone of the community and how it affects the perceived role of the facility in that particular community. Public and media relations also have responsibility for communicating the positive role the facility plays in the community, including the facility's notable achievements and the contributions it makes to the community. Although they will not likely be directly involved in conducting the risk assessments, these individuals should be intimately familiar with the results of the risk assessments. Those results should be communicated to the community in a transparent way that continues to engender trust and support from the community, while protecting the health, safety, and security of the staff and the assets of the facility.

COMMUNICATING BIORISK MANAGEMENT INFORMATION

General communication principles involve the exchange of information and opinions among risk assessors, risk managers, and other interested parties, ranging from community residents to facility staff.

The first important step is to understand that in biorisk management communication, hazards or threats are identified and communicated separately from risk. The hazard is the objects or materials present, the threat is the person or persons who may potentially cause harm to the facility, while the risk is the likelihood and consequence of a specific event occurrence involving the particular hazard or threat.

Identifying potential problems that may occur in the laboratory during both routine work and unexpected emergencies is the first step in establishing a safe and reliable laboratory. Conducting a well-designed, thorough, and routine risk assessment for the laboratory's operations provides the fundamental information to demonstrate the facility's ability to operate safely and reliably. Appropriate safety and security measures, procedures, and mitigations in place provide the necessary trustworthiness to confirm the facility's capability to conduct well-managed day-to-day operations. It is important for the laboratory to be accepted by its local community and trusted to manage its operations professionally by demonstrating reliability, honesty, and good practices. To achieve this end, results of the risk assessment must be communicated to those who work at the facility and the surrounding community. The community must believe the risk assessment results to be valid and dependable (Reid 1999). Unfortunately, in the bioscience community, risk assessments are often only an informal, even unwritten process among technical experts. The risk assessment results are rarely communicated to all staff, much less to community stakeholders.

Once the risk assessment identifies all of the current hazards and has calculated the associated risks, full cooperation of the workers and stakeholders is required in order to control or reduce the risks efficiently. This requires a good understanding of the existing risks and how to best and most effectively mitigate them to an acceptable level. The risk assessment results should provide reasons and explanation for the measures that are in place or need to be implemented. Without this knowledge, laboratory workers may see safety and security measures as disruptive requirements that have been arbitrarily implemented.

Everyone in the bioscience facility should consider it their responsibility to listen to and communicate with the larger community (Patterson et al. 2014). This type of community engagement is common in health outreach to design intervention programs and experienced by employees at large institutions that are often viewed by their communities as representatives of their institution. Transmitting accurate information to the public on hazards, risks, and the potential worst-case scenario outcomes should not be seen as an administrative function relegated to a single public information officer who relies on talking points or sound bites. Instead, the communication must be a substantive, rational consideration of shared risks with the communities in which they operate.

There is a useful corollary with a familiar numerical reference system to grade hurricane severity. The Saffir-Simpson hurricane categories rate the severity of several factors to convey the potential damage from a hurricane (National Hurricane Center 2013). On the scale, the difference between a Category 3 and Category 4 storm is the point at which people perceive a hurricane may affect them. By contrast, the method used to assign biosafety levels (BSLs) to laboratory features and the work conducted in the laboratory is not well understood by the public. Popular culture, movies, and other news media reports have sensationalized work conducted in BSL4 laboratories: they are known as the repositories for dangerous, exotic pathogens that have no cure. While this portrayal is partially true, BSL4 labs are also used to research infectious diseases that are usually life threatening. However, the likelihood and consequences portrayed in popular culture, particularly the movie industry, have been grossly exaggerated for entertainment purposes.

Not explained by these media pieces are the redundant high-containment features designed to prevent releases of pathogens into the laboratory or the environment. Work at BSL2 is so common around the world that most do not specify the safety level of the laboratory, and instead concentrate their external communication on the work conducted in the space. Simply calling out the features of the laboratory that create a high-containment or BSL4 environment would not carry the same impact as naming it a BSL4 laboratory. The NEIDL has subsequently tried to differentiate work in its BSL2 laboratories as focusing on "non-life threatening diseases like meningitis, tuberculosis, Dengue fever, and measles," and described the work in the BSL3 and BSL4 space by emphasizing the safety and security features implemented to mitigate the risk of accidental release or theft (Friday 2012). Yet the public may think that meningitis, tuberculosis, Dengue fever, and measles are dangerous diseases, and assume that the work in the BSL3 and BSL4 labs is even more dangerous.

Often communication about health risks and other scientific concerns is considered too complex for the lay public. There is a perception that the public needs the scientific community to explain complex topics, and the public should rely on the expert opinion of technocrats. Science is generally considered an open society, but open to other scientists who have obtained formal training to understand such complex topics. Inundating the public with facts in the hope that they become accustomed to the facts as presented and change their opinion does not work (Golding et al. 1992). After determining the mitigation measures that are necessary to protect against the identified risks, the institute should convey that information both to its employees and to the public.

DRIVERS OF PUBLIC PERCEPTION

It is crucial to remember that opinion based on personal experiences forms a large portion of an individual's reaction to facts, and influences how others may communicate those facts. Some may distrust the source of the facts and believe they have their own set of facts. How information makes people feel and what they hear when provided with information will substantially influence reactions more than the technical merit of the argument. Most people outside of the risk assessment community approach risk and hazards from their "gut" and their perception of the situation. A person's perception colors his or her view of reality, and if communication focuses solely on the technical facts without addressing the factors that create the audience's perception, the message will be ineffective.

Perception is driven by many factors. Among these factors are:

Control: The ability and extent to which risks may be mitigated and reduced to levels that are more acceptable.

Reversibility: This term is borrowed from developmental psychology and is a process in which individuals begin to understand logic and can process information regarding changes in their environment. It is the ability to recognize that changed situations can return to their original state.

Familiarity: How common is the hazard or threat being discussed? More common hazards and threats are more familiar, and when people have experience safely navigating the hazard or threat, such as crossing a busy street and avoiding being hit by a car, they may be willing to tolerate higher risks, as opposed to when faced with exotic hazards, such as the Ebola virus (Leiss 2004).

History: Past performance can be predictive of future results. For example, drunk drivers often repeat their behaviors.

Fairness: The concept of a hazard or threat affecting a particular subset of a population or individuals rather than being shared with the community (Rayner and Cantor 1987).

Children: People react more emotionally when children are adversely affected.

Outcome: Outcomes, particularly health-related ones, are not regarded as equal. Deaths of large numbers of people in infrequent events, such as airline crashes, or the particular risk of exposure to nuclear radiation from a nuclear power plant safety system failure, create high levels of fear in populations. Additionally, dying from cancer or other infectious or degenerative disease is more dreadful than a sudden accident (Finkel 2008).

One of the initial stages in crafting a message for the community is to listen and try to understand the perceptions, fears, and concerns in the community. Then, communicators must accentuate the level of control people may exert to address a risk. Ideally, this message will decrease the uncertainty and noise. Communication should be focused and specific, and outline the steps a person should take to mitigate the risk.

Biorisk management communication should incorporate appropriate messaging for all audiences, and the authors should consider the potential impact on psyche (the tone is as important as the content). It should address as many of the drivers of perception outlined above as possible during the risk assessment process by communicating results transparently and plainly. Then, biorisk management communication should address how specific mitigation measures control the risks, and reduce the likelihood or consequence of a catastrophic outcome.

Still, despite government agency determinations that specific high-containment labs—such as NEIDL—do not pose a high risk to the community, many people still adamantly oppose these labs being located in their communities. For example, South Boston resident and activist Klare Allen believes the 2,000-page risk assessment for NEIDL is flawed and does not prove the safety of the facility: "They have done some things, things that we've advocated for since 2002, but they haven't proven our safety and they haven't proven that the city of Boston is prepared" (Handy 2013). NEIDL failed to understand that only communicating how the risks were mitigated did not address the public concern. Instead, the communication should reflect the AMP model for biorisk management: these are the risks we evaluated, and how they are prioritized; these are the mitigation measures we have implemented to reduce those risks; and this is how we will monitor the performance of our control systems.

In the Boston example, people felt insecure and did not believe that the mitigation measures would adequately control the exotic hazards. They also felt that the risks they were concerned with were not well defined because of the general tendency to confuse the terms *hazard* and *risk* (Scheer et al. 2014). The communicators of the risk assessment failed to understand the audience perception, and what information the audience considered important, from the beginning stages of the assessment process. The time delay and insistence that the audience was misinformed, and lacked the technical understanding to fully grasp all of the security and redundant safety features, did not change the community's perceptions.

HAZARD COMMUNICATION

In most cases, in clinical diagnostic laboratory settings, the infectious agent is initially unknown. Even in the research setting, practitioners may not be very aware of hazards and threats presented by advanced techniques. Identifying potential hazards in the laboratory is one of the first steps in performing a risk assessment. The process of identifying the hazard in the laboratory needs to include information from a variety of sources. Methods to ascertain hazard information may include handling sample background information, benchmarking, conducting walkabouts and detailed inspections, holding interviews and incident reviews, performing workflow checks and process analysis, and developing a facility design revision. All of these functions require well-documented communication. How requests are presented and explained to the personnel will affect the willingness of staff to share information. Good communication around biorisks and shared interests in effective and safe laboratory work will enable assessors to obtain the data that are necessary for the risk assessment.

The risk assessment process should be dependable and reliable and have validity (Reid 1999). Confidence that the hazards and risks were assessed methodologically, and that the risk assessments are dependable, can reduce conflicts and misunderstanding of risk assessment results. Risk assessments should be conducted by a qualified team and recognized as valid by the various stakeholders.

Without the understanding and willingness of laboratory workers and others who have hands-on activities with the specific biorisks in the facility, the risk assessors cannot identify the data that address impact and probability of potential risks. As the risk assessors may not be lab workers or specialists in the area of interest, they need to gain everyone's cooperation for collecting risk assessment data. Lab personnel need to trust that the risk assessors will openly share the outcomes, allowing for timely improvement without misunderstanding or misinterpretation about the assessment results. Also, good knowledge of the source of the information will make interpretation of the risk assessment results more valid and accurate.

While the media gives attention to scientific developments and messages in an ostensible attempt to increase the public's access to scientific stories and data, their agenda may be predetermined, such as is often the case in politicizing climate change or medical stories (Mooney and Kirshenbaum 2009). Popularized by the medical disaster film *Outbreak*, a film filled with governmental secrecy and subterfuge, and the subject of alarmist news headlines when outbreaks occur, Ebola is a source of mystery that generates fear and concern—a perspective the media often emphasizes to boost their ratings. NEIDL experts dismissed the public's concern about dangerous work on deadly untreatable pathogens like Ebola because Ebola does not spread as readily among the general public like measles, but instead requires close contact to bodily fluids and contaminated objects. Hunters in the African bush, healthcare workers, and those providing care or post-mortem rituals to Ebola-infected individuals are at greatest risk of exposure. Ignoring the community fear of organ liquefaction, the NEIDL concentrated its external communications almost exclusively on the safety and security measures adopted by the laboratory to mitigate the risks of accidental and deliberate pathogen release. These mitigation measures included first aid training for high-containment staff, biometric scanners for access, bomb-resistant perimeter fencing, transportation security, and a two-man rule that prohibits scientists from working alone. They addressed the wrong risk of concern to the community.

In this instance, knowledge of the safety and security measures of the facility does not mitigate the risks as perceived by the community. The public bases its perception of the risk of Ebola on the consequences of infection. The mitigation measures cited by the laboratory affect the likelihood of such an infection occurring, but do not affect the consequences. If the NEIDL staff had spent more time listening and understanding the nature of the community's perception of risk, the NEIDL could have designed a communication strategy that would, first, acknowledge the consequences of infection and demonstrate how these consequences could only affect someone with direct exposure to the agent in bodily fluids. The communication strategy could then go on to address the likelihood of the exposure is, of course, reliant on the activity one engaged (via personal contact or contaminated articles), and emphasize the world-class healthcare facilities that could provide ameliorative care. Boston's tradition of high-quality medical care and access to world-renowned medical facilities and scientific researchers can provide supportive care and medical interventions to reduce the consequences of Ebola infection. The level of medical care in Boston far surpasses that available in Ebola endemic regions, and any infection or exposure would be managed extremely well. For example, in the 2014 outbreak, several US healthcare professionals who contracted Ebola disease after caring for infected patients in West Africa were treated in US medical facilities with advanced care. At the time of this writing, all of the infected healthcare professionals treated in US facilities have fully recovered. In contrast, the low public health capacity, poor underlying health status, reliance on traditional healers either preferentially over Western medicine or in conjunction with medical intervention, and lack of availability of personal protective equipment assist in perpetuating Ebola outbreaks in West Africa. Though a lower resourced area, South Boston does not experience all of these factors. After appropriately describing the public's risk of exposure and explaining the consequences of infection, communication can describe the methods in place to eliminate the likelihood of both exposure and infection.

Hantavirus is similar to Ebola in that both diseases are fatal because of the lack of a vaccine or treatment. Like Ebola, the only available medical intervention is supportive care. When hantavirus cardiopulmonary syndrome (HPS), caused by Sin Nombre virus, first emerged in the southwestern United States, particularly in rural areas with indigenous deer mice populations, panic set in. As the outbreak progressed, and the clinical manifestation and disease pathogenesis became better understood, critical care physicians used extracorporeal membrane oxygenation (ECMO) to reduce the 100% predictive mortality to a 67% survival and recovery rate (Wernly et al. 2011), which is only slightly better than that for Ebola. As more information on the nature of the disease transmission became available, the public health community pursued efforts to educate the population about the treatment successes as well as the mitigation measures to reduce the likelihood of exposure (e.g., wearing protection when cleaning areas with rodent droppings and reducing dust with chlorine sprays). Today in New Mexico, when a physician suspects HPS, the patient is immediately transferred to a tertiary care facility with specialists to monitor the patient's medical status and provide ECMO if the patient's condition worsens. Epidemiologists and public health nurses educate contacts and the local community on the behavioral modifications to reduce the likelihood of exposure.

COMMUNICATION OF RISK ASSESSMENT FINDINGS

Work with pathogens will always involve some level of risk, and the potential for human error always exists. Thus, institutes must specifically communicate the outcomes of risk assessments to their employees to help those workers understand the risks in their own work environment, and how they can efficiently and effectively reduce those risks on a daily basis without disrupting the objectives of their work. Risk assessment results, regardless of whether quantitative or qualitative, will provide some reference point that will help people understand the level of risk associated with each of their own activities. The results should also help in explaining why certain mitigation measures are necessary to reduce the risks, and why the performance of those measures needs to be monitored.

Because of uncertainties or insufficient scientific data when the identity of hazards in samples is unknown, risk assessments can be based on incomplete knowledge or information. Biological risk assessment in such a situation is mostly a subjective process that involves professional judgments; there will be inherent limitations and assumptions made in the process. Risk communication is even more important when the risk assessment relies on primarily subjective analysis. Risk communication should assist all stakeholders—from executive management to laboratory technicians—to understand how the risks in the laboratory have been identified, analyzed, and prioritized, and the affected staff need to embrace the resulting risk management decisions. Again, as with all risk communication, this interchange must be multidirectional, allowing all of the stakeholders to contribute to the risk assessment process.

Even if the risk assessment depends on subjective interpretation, it should always be both formal and documented. If the risk assessment is not formally conducted, and the risks are not clearly communicated to the staff, the staff may not appreciate all of the risks that they contribute to the lab environment on a daily basis, and they will likely resist using any mitigation measure that introduces inconvenience into the system without substantive justification. Others may comply with the administrative requirements, but may do so without understanding the benefit of such processes half-heartedly or not believing the rationale behind the decisions. Action without acceptance of the rationale and overt resistance reflect the absence of a durable biorisk management culture, and hazard and risk communication arguably sits at the center of this problem.

Unfortunately, traditional biosafety messages tend to focus on only specific safety mitigation measures, such as using a biosafety cabinet or donning and doffing personal protective equipment. These messages also tend to be relatively generic: all of the employees in a certain facility, or even everyone who attends training from multiple institutes, are taught the same safety procedures. In a biorisk management culture, by contrast, the communication with the staff of an institute should emphasize the risks that are unique to an individual laboratory, explain the prioritization of those risks, and then show how specific mitigation measures will reduce those risks to a level that management deems acceptable. The messages and the target audience would constantly vary depending on the specific work, the people involved in that work, and the location of that work.

In addition to transparent, internal communication about risk assessment, an institute has an obligation to communicate risk assessment outcomes to the external community. This communication could be preceded by a period of listening to the perspectives and perceptions of the community, such as focus groups. Risk assessment results should be communicated to the community within the context of the community's concerns. Since there will always be some level of risk for a bioscience facility that works with pathogens, care should be taken in how those residual risks are communicated to the public. The communication should include both how the facility reduces the likelihood of those risks happening and how the consequences can be managed if an incident occurs. Most importantly, the communication should demonstrate the facility's stewardship of safety and security, and its commitment to the health and well-being of its community. Managers and communicators must carefully listen to how the facility's risk assessment message is received by the community, and be prepared to alter or augment the facility's mitigation measures based on feedback from the community. Risk communication is not persuasion, advertisement, or propaganda; rather, it is process designed to strengthen risk management and demonstrate that the facility is a valuable and responsible member of its local community and the broader society.

RISK COMMUNICATION IN A RISK GOVERNANCE FRAMEWORK

In 2005, the International Risk Governance Council (IRGC), an independent non-profit organization focused on improved understanding of systematic risk, established a risk governance framework (International Risk Governance Council 2005), which identifies four phases of risk governance: preassessment, risk appraisal (assessment), tolerability and acceptance judgment, and risk management. The IRGC emphasizes that risk communication is an essential element of each of these phases. Although IRGC uses somewhat different terminology, these four phases are directly analogous to the elements of the risk assessment process of a biorisk management system: hazard identification, risk assessment, risk tolerance and acceptability, and risk mitigation and evaluation. The following sections describe the communication that should occur during each phase of the risk governance framework.

PREASSESSMENT (HAZARD IDENTIFICATION)

The individual or team responsible for conducting the risk assessments must communicate frequently and substantively with those at the facility who work directly with hazards, including the principal investigators/scientists/researchers, contractors and custodial staff, and laboratory managers. In addition, the risk assessment team may need to communicate regularly with the administrative or technical personnel who have responsibility for facility inventories, and the security personnel who have responsibility for the facility access controls. The risk assessment team should collect all relevant information about the hazards in the facility, how those hazards are used and stored, who has access to them, and what procedures are in place. At the same time, the risk assessment team must clearly explain the objectives and methodology of the risk assessment, engaging the stakeholders in the intellectual rationale for the activity. A close communication link between risk assessors, managers, and workers, particularly in this phase, is crucial for improving overall governance and performance based on mutual trust of each other.

Once the risk assessment team has identified the hazards, they must help to establish a communication vehicle for transparently describing those hazards. For example, a biological agent material safety data sheet (MSDS) describes the generic characteristics of that particular agent; everyone who works with or has access to a particular agent should also be familiar with the MSDS for that agent. Although the MSDS characterizes a hazard, the MSDS is not a risk assessment.

During this phase, the risk assessment and risk communication teams should also be listening to those at the facility who do not work directly with the hazards, including executive management and the administrative and technical staff, as well as the local community. The risk assessment and risk communication teams should endeavor to understand the extent of knowledge about the facility's hazards, and the perceptions of the risks involved with working with those hazards. Understanding the opinions and concerns of those who do not work directly with the hazards is another essential data gathering and data communication activity that must inform the risk assessment process.

RISK ASSESSMENT

Once the hazards have been identified, the risk assessment team must ask a simple question: "What can go wrong?" Answering that question reveals the risks. Risks involve not only hazards, but also the particular work with the hazards, the people who conduct the work, the equipment used for the work, and the location of the work. Communication between risk assessors and laboratory workers and managers is essential to identify all the risks. Once all the risks are identified, the risk assessment team must analyze the likelihood and consequences of each of the individual risks, and then prioritize those risks. Again, this risk assessment cannot be conducted in a vacuum, and must involve clear communication among those with expertise in risk assessment, those with intimate knowledge of the hazards and the specific work activity of the laboratory, and those familiar with the engineered, administrative, and procedural controls. Because a single meeting is generally insufficient for all of these stakeholders to participate fully in the risk assessment process, transparent communication is the glue that holds the risk assessment phase together. A risk assessment that does not successfully engage all of the relevant stakeholders is a failure that will significantly impact subsequent phases of the risk governance process.

During this phase, the risk assessment and risk communication teams should be listening to and communicating with those at the facility who do not work directly with the hazards, including executive management and the administrative and technical staff, as well as the local community. The risk assessment and risk communication teams need to be acutely aware of any change in sentiment, either internal or external, regarding the work at the facility, and how that might affect the technical risk assessment process. In addition, it is critical for those not directly involved with the hazardous work of the facility to understand that the facility exercises a rigorous and intellectually sound risk assessment system. The risk assessment methodology itself should be shared with these stakeholders.

TOLERABILITY AND ACCEPTANCE JUDGMENT (RISK TOLERANCE AND ACCEPTABILITY)

The most important outcome or result from a risk assessment is the prioritization of the risks. In general, a risk assessment that considers only one risk is much more difficult to interpret than one that compares multiple risks against each other. A single risk of moderate does not mean very much to those asked to review that risk assessment, but a single risk of moderate when all the other risks in that laboratory are either low or high suddenly becomes much more significant. In most cases, work with one pathogen in one laboratory will involve multiple risks, and the risk assessment for that laboratory should reflect that reality.

During this phase, the risk assessment team must communicate the prioritization of the risks to the laboratory directors and executive management, clearly explaining why certain risks have been judged higher than others. The laboratory directors and executive management must use this information to decide which risks are acceptable and which are not acceptable. A risk-tolerant management team will accept more risk than a risk-adverse management team—even if the risks are exactly the same. A high risk is not necessarily unacceptable, and a low risk is not necessarily acceptable. At this point, communication between the risk assessment team and management must focus on whether the implementation of additional control measures could reduce the unacceptable risks to an acceptable level. Additional experts, such as those with knowledge of the technical work and the various safety and security measures, should be engaged in the discussion. The management team should decide what additional control measures to implement for which specific risks, and how the performance of those controls will be measured. With these new control measures in place, management should believe that the previously unacceptable risks are now acceptable. If unacceptable risks remain, the work should not proceed. The final acceptable risks, including the respective mitigation measures and their performance metrics, should be clearly documented under management's signature.

The key step in this phase is to communicate management's risk acceptance and required mitigation and performance measures to all of the relevant stakeholders. Ideally, the communication between management, the risk assessment team, and the relevant stakeholders has been frequent and transparent so that these decisions are well understood. Regardless, those who work directly with the hazards must fully accept the rationale for the use of the specified risk mitigation measures (Hance et al. 1988; Lundgren 1994). Effective risk communication will enable an environment of trust for assessing and mitigating risks and related concerns within the institute, will foster tolerance for conflicting viewpoints, and provide the basis for their resolution.

During this phase, the risk assessment and risk communication teams again should be listening to and communicating with those at the facility who do not work directly with the hazards, including the administrative and technical staff, as well as the local community. The risk assessment and risk communication teams need to be acutely aware of any change in sentiment, either internal or external, regarding the work at the facility, and how that might affect the technical risk assessment process. Based on this information, the risk assessment and risk communication teams must determine how best to communicate the results of the risk assessment to the local community. This particular external communication is a fundamental element of a well-functioning biorisk management system.

Risk Management (Risk Mitigation and Evaluation)

The risk management phase involves the implementation of the designated risk mitigation measures, and the continuous evaluation of the performance of those mitigation measures. It is the responsibility of everyone involved in the activities to ensure that the mitigation measures and performance metrics are used as intended in the risk assessment. Management must ensure that they communicate the risks, the risk assessment, the control measures, and the performance metrics to anyone new to the particular environment through training or other means. Even regular employees will need to be regularly retrained to ensure that they remain familiar with the entirety of the biorisk management system.

In addition to providing regular training for the staff, the risk assessment team— as well as the laboratory directors, scientists, and technicians—must remain acutely aware of any changes to the research protocols, the equipment, or any other factor that could alter the risk assessment. Risk communication must continue throughout a research project so that the risk assessment can be revised whenever necessary.

Perhaps the most important role of internal risk communication during this phase revolves around the evaluation of the control measures. Many different stakeholders should be involved in monitoring the performance of the mitigation measures; those roles and responsibilities should be clearly articulated before the activity begins. It is critical that everyone involved in the work understands the specific performance metrics, and how those data will be collected throughout the duration of the project. The results of those performance evaluations need to be shared transparently and regularly with all of the relevant stakeholders. A communication protocol should be established so that mitigation measures that do not perform as anticipated can be modified immediately, or the risk assessment can be revised.

During this phase, the risk assessment and risk communication teams again should be listening to and communicating with those at the facility who do not work directly with the hazards, including the administrative and technical staff, as well as the local community. The risk assessment and risk communication teams need to be acutely aware of any change in sentiment, either internal or external, regarding the work at the facility, and how that might affect the performance of the designated control measures. Based on this information, the risk assessment and risk communication teams must determine how best to communicate any results from the performance monitoring of the risk mitigation measures.

CRISIS COMMUNICATION

The largest difference between risk and crisis communication is that in a crisis, information is incomplete and evolving. Additionally, during crisis communication, human behavior and reaction to information are greatly affected by noise—or superfluous information. The concept of noise disrupts cognitive function when individuals are focused on basic needs, such as physiological (air, food, water, shelter), and safety and security (Maslow 1943). Noise can increase risk if those who receive a message perceive a threat to their personal well-being. Alternatively, if a message reassures them that their basic needs will be met and they will be safe and secure, it is much more likely that they will select an appropriate response to the situation (Slovic 1980).

Most institutions fail to prepare for crisis communication because they fail to consider worst-case scenarios. There is often overconfidence or complacency that the system will function as engineered, and that it has been engineered to withstand a catastrophic system failure. Most institutions do not have emergency communication plans that articulate how to respond to a catastrophic system failure. Although it is human nature to focus least on events that are extremely unlikely, a substantive risk assessment process will compel an institute to consider everything that can go wrong, including low-probability events.

In a crisis, messages need to be distilled down to the most relevant information transmitted in the simplest and most timely manner. The timeliness of initial communication often reflects the level of preparedness and anticipation of unintended consequences. The more rigorous a risk assessment, the more likely a facility will be able to identify potential vulnerabilities early and prepare for potentially catastrophic events. In general, an institute that has established a reputation for dependable risk communication will transition much more smoothly and successfully into crisis communication than an institute that has a poor track record in risk communication. Institutions can measure their performance and preparedness by using tools such as annual drills and establishing a way to collect feedback. Monitoring the performance of the communications is important to guide and adjust future messages (Johansson and Harenstam 2013).

Crisis communicators—often the institution's leadership—must demonstrate empathy, explain the known facts, acknowledge uncertainty, and clarify the steps being taken to mitigate the adverse outcomes. They should also focus on what people can control, and provide them with clear guidance. Frequent, well-prepared updates can develop trust between the communicator and the audience, and can help control rumors and manage expectations. Obviously, poorly prepared messaging can have a counterproductive effect.

Risk and Crisis Communication during Hurricane Katrina

Hurricane Katrina, the deadliest Atlantic tropical cyclone during the 2005 season, which caused US$108 billion in damage (National Oceanic and Atmospheric Administration 2005), is a good example of where poor risk assessment led to poor risk communication, which, in turn, led to disastrous crisis communication. The risks associated with the hurricane protection system in New Orleans, Louisiana, were never quantified through a rigorous risk assessment process. No one considered whether the levee system would breach at the points it did. In fact, the failures in the levee system occurred where they were least expected even though the post-event investigation identified inadequate design and construction as the primary cause of the flooding. One particular risk of catastrophic levee system failure that was not included in the risk assessment was loss of life.

Because of an inadequate assessment of the risks of the levee system, the risks of the levee system failing were never communicated to the public. As a result, those who lived in the shadow of the levee system were not aware that system failure might lead to death. They believed the levees and seas walls would protect them from hurricanes and floods. They relied on historical experiences to make decisions despite the government's instructions for preparedness and evacuation (Elder et al. 2007). This behavior is similar to interviews with individuals who survived other hurricanes and their decision-making process. Those individuals found the government's warnings not to be credible because they had lived through many hurricanes in the past (Tierney et al. 1999).

Poor risk communication in New Orleans doomed the government's ability to communicate during the crisis. The public had little confidence in the competency or knowledge of the government officials who had responsibility for crisis communication (Cordasco et al. 2007). Some perceived the warnings about the severity of the storm as scare tactics, and some feared accidents on the highway more than "a little water." Additionally, the warnings to evacuate did not use the word *mandatory*. Many people interviewed after Katrina mentioned that the message did not convey the full risk posed by the storm; instead, it was a mixed message that anticipated a Category 5 storm, but included "if you can leave, please leave."

In 2007, the American Society of Civil Engineers released a report about Katrina entitled "What Went Wrong and Why." The report presented a series of recommendations, including the creation of a rigorous risk assessment process to quantify the risks for the populations, and a substantive public risk communication plan. The report asserted that the state and local governments should aim to create an informed and engaged public, which means ensuring that the public is cognizant of the risks associated with worst-case scenarios (American Society of Civil Engineers 2007).

CONCLUSION

The Hurricane Katrina example demonstrates that the greatest communication failures usually happen during crises. For the NEIDL in Boston, the crisis started when the public organized a campaign to prevent the NEIDL from working with dangerous pathogens—well after building construction had begun, scientists and staff had been hired, and hundreds of millions of US dollars had been committed to the operation of the facility. But because the NEIDL's proponents had not considered this risk in their risk assessments, and had not developed a sound, proactive risk communication plan, the NEIDL was poorly prepared for the crisis that ensued.

As this chapter discussed, risk communication must be based on and is required to facilitate a comprehensive risk assessment, which includes a consideration of everything that could go wrong. Part of that risk assessment must consider the risk perception in the community. Risk assessors and communicators must engage all stakeholders, including the public, early in this process, endeavoring to listen and understand everyone's concerns and perspectives. A relationship must be built with all stakeholders before, during, and after the risk assessment and mitigation process. Risk communication will be much more successful if an institute can include the public in many elements of the biorisk management system. Establishing transparent, positive, risk-based communication with the public before major or catastrophic events occur is one of the best ways to mitigate potential adverse reactions in the event of a crisis.

REFERENCES

American Society of Civil Engineers. 2007. The New Orleans Hurricane Protection System: What Went Wrong and Why. http://www.asce.org/uploadedfiles/publications/asce_news/2009/04_april/erpreport.pdf (accessed October 2, 2014).

Beecher, N., E. Harrison, N. Goldstein, M. McDaniel, P. Field, and L. Susskind. 2005. Risk Perception, Risk Communication, and Stakeholder Involvement for Biosolids Management and Research. *Journal of Environmental Quality*, 34(1): 122–128.

Boston University Medical Campus National Emerging Infectious Diseases Laboratory. n.d. Frequently Asked Questions. http://www.bu.edu/neidl/resources/faq/ (accessed September 10, 2014).

Boston University Medical Campus National Emerging Infectious Diseases Laboratory. 2014. Statement on Classified Research. http://www.bu.edu/neidl/research/no-classified-research/ (accessed August 14, 2014).

Cedrone, A. 2012. Residents Oppose BU Biolab for Deadly Diseases in South End. *Boston Globe*, April 20. http://www.bu.edu/neidl/2012/04/20/residents-oppose-bu-biolab-for-deadly-diseases-in-south-end/.

Cordasco, K.M., D.P. Eisenman, D.C. Glik, J.F. Golden, and S.M. Asch. 2007. "They Blew the Levee": Distrust of Authorities among Hurricane Katrina Evacuees. *Journal of Healthcare for the Poor and Underserved*, 2007: 277–282.

Elder, K., S. Xirasagar, N. Miller, S.A. Bowen, S. Glover, and C. Piper. 2007. African Americans' Decisions Not to Evacuate New Orleans before Hurricane Katrina: A Qualitative Study. *American Journal of Public Health*, 2007: S124–S129.

Finkel, A.M. 2008. Perceiving Others' Perceptions of Risk. *Annals of the New York Academies of Sciences*, 2008: 121–137.

Friday, L. 2012. NEIDL Goes Public. *BU Today*, January 26. http://www.bu.edu/today/2012/neidl-goes-public/ (accessed June 03, 2014).

Golding, D., S. Krimsky, and A. Plough. 1992. Evaluating Risk Communication: Narrative vs Technical Presentations of Information about Radon. *Risk Analysis*, 1992: 27–35.

Hance, B.J., C. Chess, and P.M. Sandman. 1988. *Improving Dialogue with Communities: A Risk Communication*. New Brunswick, NJ: Rutgers University, Environmental Communication Research Programme.

Handy, D. 2013. Residents Continue to Fight against BU Infectious Disease Lab. Boston NPR. http://www.wbur.org/2013/04/11/infectious-disease-lab-fight (posted April 11, 2013).

International Risk Governance Council. 2005. *IRGC White Paper No. 1: Risk Governance—Toward an Integrative Approach*. Geneva. http://www.irgc.org/IMG/pdf/IRGC_WP_No_1_Risk_Governance__reprinted_version_.pdf.

Johansson, A., and M. Harenstam. 2013. Knowledge Communication: A Key to Successful Crisis Managment. *Biosecurity and Bioterrorism: Biodefense Strategy, Practice, and Science*, 11: s260–s263.

Leiss, W. 2004. Effective Risk Communication Practice. *Toxicology Letters*, 149(1-3): 399–404.

Lundgren, R.E. 1994. *Risk Communication: A Handbook for Communicating Environmental, Safety, and Health Risks*. Columbus, OH: Battelle Press.

Maslow, A.H. 1943. A theory of human motivation. *Psychology Review*, 50(4): 370–396.

Mooney, C., and S. Kirshenbaum. 2009. Unpopular Science. *The Nation*, August. http://www.thenation.com/article/unpopular-science?page = full.

National Academies. 2010. *Continuing Assistance to the National Institutes of Health on Preparation of Additional Risk Assessments for the Boston University NEIDL, Phase I.* Washington, DC: National Academies Press.

National Hurricane Center. 2013. Saffir-Simpson Hurricane Wind Scale. May 24. http://www.nhc.noaa.gov/aboutsshws.php.

National Institute of Allergy and Infectious Diseases. 2004. NIAID Biodefense Research Agenda for Category B and C Priority Pathogens. Progress Report. National Institutes of Health.

National Institute of Allergy and Infectious Diseases, NIH. "Final Environmental Impact Statement National Emergency Infectious Diseases Laboratory." Boston University Medical Campus National Emerging Infectious Diseases Laboratory. December 2005.

National Institute of Allergy and Infectious Diseases. 2007. NIAID Strategic Plan for Biodefense Research. National Institutes of Health.

National Institute of Allergy and Infectious Diseases. 2010. The Need for Biosafety Laboratory Facilities. Biodefense and Emerging Infectious Diseases. July 23. http://www.niaid.nih.gov/topics/biodefenserelated/biodefense/publicmedia/pages/biosafetylabfacs.aspx (accessed June 04, 2014).

National Oceanic and Atmospheric Administration. 2005. Hurricane Katrina. Extreme Events Special Reports. December 29. http://www.ncdc.noaa.gov/extremeevents/specialreports/Hurricane-Katrina.pdf.

Patterson, A., K. Fennington, R. Bayha, D. Wax, R. Hirschberg, N. Boyd, and M. Kurilla. 2014. Biocontainment Laboratory Risk Assessment: Perspectives and Considerations. *Pathogens and Disease*, 71(2): 1–7.

Rayner, S., and R. Cantor. 1987. How Fair Is Safe Enough? The Cultural Approach to Societal Technology Choice. *Risk Analysis*, 7(1): 3–9.

Reid, S.G. 1999. Perception and Communication of Risk, and the Importance of Dependability. *Structural Safety*, 21(4): 373–384.

Scheer, D., C. Beninghaus, L. Beninghaus, O. Renn, S. Gold, B. Roder, and G.F. Bol. 2014. The Distinction between Risk and Hazard: Understanding and Use in Stakeholder Communication. *Risk Analysis*, 34(7): 1270–1285.

Slovic, P. 1980. Facts and Fears: Understanding Perceived Risk. In *Societal Risk Assessment*, ed. R.C. Schwing and W.A. Albers Jr. New York: Springer, pp. 181–216.

Tierney, K.J., M.K. Lindell, and R. Perry. 1999. *Facing the Unexpected: Disaster Preparedness and Response in the United States.* Washington, DC: Joseph Henry Press.

Wernly, J.A., C.A. Dietl, C.E. Tabe, S.B. Pett, C. Crandall, K. Milligan, and M.R. Crowley. 2011. Extracorporeal Membrane Oxygenation Support Improves Survival of Patients with Hantavirus Cariopulmonary Syndrome Refractory to Medical Treatment. *European Journal of Cardio-Thoracic Surgery*, 40(6): 1334–1340.

10 Three Recent Case Studies
The Role of Biorisk Management

Reynolds M. Salerno

ABSTRACT

This chapter reviews three events that occurred at the US Centers for Disease Control and Prevention and the US National Institutes of Health in 2014. This chapter then analyzes how biorisk management systems could reduce the frequency of these types of events occurring in the future.

INTRODUCTION

Three high-profile events took place during the first half of 2014. These events occurred at the US Centers for Disease Control and Prevention and the US National Institutes of Health, the two US government organizations that publish *Biosafety in Microbiological and Biomedical Laboratories* (BMBL), the document that most clearly articulates the current paradigm for biosafety in the United States (US Department of Health and Human Services 2009). As the introduction to this book explains, the United States bioscience community has been slow to embrace the concept of biorisk management, which is fundamentally different than the views espoused by the BMBL.

This chapter reviews these three events, and the US government's response to them. In addition, this chapter considers what elements of biorisk management may have been inadequately practiced or wholly absent, and suggests that comprehensive biorisk management systems may have either prevented these incidents or at least identified them much sooner.

We recognize that hindsight allows us to see things that were not obvious at the time. But we believe that it is important for the bioscience community to carefully evaluate choices that have been made in the past, and to consider if different management decisions could have led to different outcomes.

CASE STUDY 1: ANTHRAX MISHANDLING AT THE CDC

In June 2014, the CDC's Bioterrorism Rapid Response and Advanced Technology (BRRAT) biosafety level 3 (BSL3) laboratory transferred *Bacillus anthracis* extract to the biosafety level 2 (BSL2) laboratories in the Bacterial Special Pathogens Branch (BSPB) and the Biotechnology Core Facility Branch (BCFB). Following this transfer, a scientist in the BRRAT laboratory determined that the *B. anthracis* that had been transferred out of the BSL3 laboratory had not been completely inactivated. The incident was immediately reported to the CDC's Environment, Safety, and Health Compliance Office (ESHCO) and the CDC's Division of Select Agents and Toxins (DSAT). Eventually, after 11 days, ESHCO determined that at least 67 CDC staff members and three CDC visitors may have been unwittingly exposed to viable anthrax cells or spores, and all reported to the CDC clinic and were placed on post-exposure prophylaxis with antibiotics and vaccine (US Centers for Disease Control and Prevention 2014a; Russ and Steenhuysen 2014).

Although it seems that no staff members became ill with anthrax, the official CDC report on the incident concluded that "this was a serious and unacceptable incident which should have never happened" (US Centers for Disease Control and Prevention 2014a). The work of the BRRAT laboratory was immediately suspended, and a moratorium was placed on the transfer of any biological material from a CDC BSL3 or BSL4 laboratory until improved measures could be implemented. In early July, contentious congressional hearings were held, and later that month Michael Farrell, director of the BRRAT laboratory, who had been reassigned in June, resigned from the CDC (Morgan 2014).

A ROLE FOR BIORISK MANAGEMENT?

The official CDC report on this incident states that "the overriding factor contributing to this incident was the lack of an approved, written study plan reviewed by senior staff or scientific leadership to ensure that the research design was appropriate and met all laboratory safety requirements" (US Centers for Disease Control and Prevention 2014a). In the language of biorisk management, this particular experiment lacked a comprehensive, documented risk assessment.

The concluding section of the official CDC report indicates that in the future, laboratories across the CDC should "use an approach that identifies the points in any project where potential mistakes would have the most serious consequences [and] that provides specific actions to avoid these mistakes." In the language of biorisk management, this is exactly what an experiment-specific or activity-level risk assessment should do. The failure of one of the most prestigious bioscience facilities in the world to systematically conduct risk assessments for its work reflects a deficiency in the current paradigm that is based on biosafety levels and biosecurity regulations.

As Chapter 3 explains, every experiment should have its own risk assessment that articulates everything that could go wrong (risks), and evaluates the likelihood and consequences of all of those risks. That document should be developed by those with the most knowledge about the work, as well as those who are not personally or professionally invested in the activity itself. The risk assessment should be reviewed by an independent peer organization, such as an institutional biosafety committee, and approved by management. It seems that there was a lack of even an informal risk assessment process in the operations of this particular laboratory at the CDC.

Another fundamental aspect of any biorisk assessment should be documented familiarity with the peer-reviewed literature in the field under study. According to the CDC report, "a review of the literature would have found that filtration has been recommended for inactivation of *B. anthracis*.... Drevinek et al. (2012) concluded that the formic acid method (as used by the BRRAT laboratory) did not sterilize *B. anthracis*; they also used centrifugal filtration to remove viable particles (including spores) from *B. anthracis* preparations" (US Centers for Disease Control and Prevention 2014a). Lack of familiarity with the scientific literature represents a significant flaw in the risk assessment and management process.

Upon completion of a risk assessment, one of the first questions that should be asked is: Could the hazard be eliminated from this experiment, or could it be substituted for a lesser hazard? In this case, it seems that neither of these fundamental questions was ever considered. According to the CDC report,

> The BRRAT laboratory supervisor instructed the laboratory scientist to use virulent strains because of the possibility that avirulent strains might not yield the same MALDI-TOF profile. However, the instrument manufacturer states that the system identifies bacteria to only the species level and would not distinguish strains of the same species. The use of avirulent strains to develop protocols would have been appropriate, particularly when conducting a pilot study (US Centers for Disease Control and Prevention 2014a).

In fact, the concluding section of the CDC report suggests that, in the future, laboratories across the CDC should "promote the use of non-pathogenic organisms in research and training activities, whenever possible." It is clear that experts at the CDC understand that elimination and substitution are the most effective biosafety risk mitigation measures, but the system in which they operate does not seem to prioritize this perspective.

As Chapter 5 explains, practices and procedures for every experiment should flow directly from the risk assessment process. In other words, the procedures should be explicitly designed to be part of the risk mitigation for the identified risk of that particular work. In this case, the BRRAT laboratory relied on a procedure that had been used by the BSPB laboratory for *Brucella* species. Although all of the BRRAT laboratory's procedures with *B. anthracis* included a filtration step, in this case the BRRAT lab did not filter extracts because it was not part of the BSPB *Brucella* protocol. According to the CDC report, "since *B. anthracis* forms spores that are more resistant to inactivation by chemicals than vegetative cells, the BRRAT laboratory scientist's assumption that the same treatment would apply to *B. anthracis* was incorrect.... The BRRAT laboratory's SOP for assuring sterility was specific for DNA preparations, and SOPs for other materials do not appear to have been in place" (US Centers for Disease Control and Prevention 2014a).

In addition, the BRRAT laboratory did not have written, approved procedures to reliably ensure that organisms were no longer viable before removing them from

BSL3 containment. Although the lab SOP for DNA preparations indicated that sterility check plates for *B. anthracis* should remain in incubation for up to 48 hours, a BRRAT scientist modified the methods from the BSPB laboratory and shortened the incubation period to 24 hours. The growth of *B. anthracis* on the sterility plates took place in the incubator sometime between day 1 and day 8.

One of the most striking aspects of this incident is that it was only accidentally discovered. After the 16 sterility plates had been incubated for 24 hours, and no growth had been observed, the laboratory scientist planned to autoclave the plates and then discard them. "However, the individual had difficulty opening the autoclave door. As a result, the plates were returned to the incubator and left for seven additional days…. If the plates had been autoclaved after 24 hours, as planned, the event would never have been discovered" (US Centers for Disease Control and Prevention 2014a).

As Chapter 8 explains, a critical element of a well-functioning biorisk management system—but widely disregarded in most biosafety laboratories—is an explicit process for continually evaluating the performance of the existing mitigation measures. The risk assessment should be the basis for establishing performance metrics for the risk mitigation measures in place for specific experiments or activities. According to the CDC report, "a written protocol to certify the sterility of the material to be transferred to BSL2 laboratories was not in place, and the BSL2 laboratories did not have an SOP that required receipt of written certification of non-viability for transfers prior to acceptance of microbiological material" (US Centers for Disease Control and Prevention 2014a). In this case, there should have been a specific testing protocol to demonstrate inactivation prior to any transfer of materials outside of the BSL3 laboratory. Such a performance metric would effectively help to mitigate the risk of exposing individuals to the live agents outside of containment.

The conclusion of the CDC report identified the following conditions that allowed this event to occur: "failures of policy, training, scientific knowledge, supervision, and judgment on the part of this laboratory" (US Centers for Disease Control and Prevention 2014a). Lack of oversight and training seems to be particularly evident. The scientist in the BRRAT laboratory had not performed this specific procedure with a pathogenic agent before, and the scientist had no previous experience transferring select agent-derived materials, other than DNA preparations, from BSL3 to BSL2 laboratories. The scientist was not familiar with the current literature, and presumably, neither was the supervisor. The scientist should not have been instructed to proceed without submitting a complete protocol for review and approval. And such approval should not have been granted until the scientist had completed requisite training that demonstrated adequate knowledge of the documented procedures.

Although the CDC report implies that the primary blame for this incident lies with the BRRAT laboratory, it is clear that the absence of institutional policies, training, and oversight also contributed to this incident. For example, the report indicates that there is a CDC policy that requires each individual to swipe his or her own identification card key before entering a secured area, but acknowledges that there is a "practice of authorized staff piggy backing" or following an authorized colleague into a secured area. This practice, which management is responsible for failing to eliminate, significantly delayed the ability of CDC responders to determine who had been potentially exposed to the anthrax.

As Chapter 6 discusses, an institution's training program should be an explicit risk mitigation function. In other words, a management system should be in place that translates the outcomes of the risk assessment into an activity-specific and continuous training program. One of the key aspects of designing a training program is understanding and documenting who needs training, and their current knowledge, skills, and abilities. Then, based on an analysis of the gaps between current knowledge and required knowledge to reduce the identified risks, a training program can be designed and implemented. Although training is one of the most common activities in the biosafety community, it is generally viewed as an activity that is independent of specific experiments or laboratory activities. The biosafety community largely relies on generalized training; training is rarely designed to mitigate risks identified in an experiment-specific risk assessment.

The official CDC report also identified a failure in communications, especially as the incident unfolded. On the day that the incident was first identified, two CDC staff members went to the emergency department at Emory University. Eventually, after many days of event investigation, as many as 84 people were identified as potentially exposed. The staff members of the affected laboratories, as well as the biosafety response teams, were reluctant to communicate anything about the event even to their CDC colleagues. According to the CDC report, "CDC scientists who worked near the impacted laboratories commented that they first learned of the event by witnessing CDC closing and/or decontaminating laboratories rather than through direct communication" (US Centers for Disease Control and Prevention 2014a). Once it became clear that a potential widespread anthrax exposure had occurred, "the CDC clinic was overwhelmed at times during the response" (US Centers for Disease Control and Prevention 2014a). The failure to quickly disclose the details of the incident also led to inconsistencies in the decontamination methods to minimize the consequences of the exposure. The CDC report acknowledges that "there was no clear overall lead for the incident in the first week," and that "in retrospect, it is clear that broad communications should have occurred earlier in the process" (US Centers for Disease Control and Prevention 2014a). The failure to communicate effectively about the incident contributed to its escalation into a crisis.

As Chapter 9 discusses, risk communication is a critical element of a biorisk management system. In particular, risk communication should be understood as another component of the risk mitigation process. In this case, a substantive risk assessment would have considered worst-case scenarios, and identified the risk of transferring pathogenic materials into low-containment laboratories without complete inactivation. A risk communication plan would consist of communicating this risk to everyone even remotely involved with this work. In addition, that plan would determine in advance that immediate, transparent communications would best facilitate the response, as well as minimize confusion and fear, by detailing clear roles and responsibilities. As Chapter 9 explains, an institute that has established a reputation for dependable risk communication will transition much more smoothly and successfully into crisis communication than an institute that has a poor track record in risk communication.

Ultimately, this incident at the CDC was a failure in institutional biorisk management leadership. Although the blame has been assigned to one laboratory, and specifically the director of that individual laboratory, it seems clear that many of the failures were systemic, and not attributable to one single person or one single laboratory. It does not appear that the CDC had a broad-based biorisk management system in place at the time of this incident. Instead, as is quite common in the bioscience community, the CDC apparently relied on the paradigm of biosafety levels and biosecurity regulations to manage its risks. Soon after the incident, CDC Director Thomas Frieden was asked by a reporter about how closely the CDC follows the CWA 15793 guidance on laboratory biorisk management. Frieden responded, "Those guidelines are out there globally. They are not necessarily the most appropriate, useful, or protective for US laboratories." Instead, he indicated that the United States relies on many regulations that govern how dangerous pathogens are handled (Russ and Steenhuysen 2014; Steenhuysen and Begley 2014). However, a biorisk management system should be designed to supplement laboratory operations that otherwise rely primarily on compliance with national regulations.

CASE STUDY 2: H5N1 INFLUENZA MISHANDLING AT CDC

In January 2014, a laboratory in the CDC's Virology Surveillance and Diagnosis Branch (VSDB) unintentionally cross-contaminated a culture of low pathogenic avian influenza A (H9N2) with a strain of highly pathogenic avian influenza A (H5N1). CDC shipped an aliquot of the contaminated H9N2 virus to the US Department of Agriculture (USDA) Southeast Poultry Research Laboratories (SEPRL) in March 2014. Because the VSDB influenza laboratory was unaware of the contamination, the material was shipped to SEPRL as Category B biological substances—standard shipping procedures for some infectious agents—but not as Category A infectious substances. Appropriate select agent transfer procedures, including permitting, notifications, and safety precautions, were not followed. Subsequently, some of the contaminated culture was transferred to other CDC influenza laboratories (US Centers for Disease Control and Prevention 2014b).

In May 2014, SERPL informed the CDC that the materials that they received were contaminated with highly pathogenic H5N1 influenza, and the VSDB influenza laboratory immediately confirmed the mistake. Because the work with the contaminated materials at both the CDC and SEPRL took place inside BSL3 facilities, it is believed that no staff members were exposed to the highly pathogenic H5N1 influenza. By the end of June, all contaminated H9N2 stocks had been destroyed or were secured in freezers approved to store select agents. Nevertheless, the CDC supervisory chain of command, including CDC leadership, was not informed of this incident for almost three more weeks (US Centers for Disease Control and Prevention 2014b).

A ROLE FOR BIORISK MANAGEMENT?

The official CDC report on this incident focuses on the technical details of the work in the VSDB laboratory, where the cross-contamination occurred. The report does not indicate whether a formal risk assessment had been documented for this particular work, or whether that risk assessment had identified the risk of cross-contamination (US Centers for Disease Control and Prevention 2014b). Based on the events that occurred, however, either the risk of cross-contamination was not articulated in the risk assessment, the mitigation measures to address this risk were not put in place or validated to be performing as designed, the mitigation measures were not followed, or a formal risk assessment was not conducted. Regardless, the incident has its origin in the CDC risk assessment process.

As Chapter 3 explains, every experiment or work activity deserves its own documented risk assessment, which describes what could go wrong. It is fair to assume that a laboratory that works with both low pathogenic and high pathogenic subtypes of avian influenza would be concerned about cross-contamination. Was that risk identified, and what control measures were established specifically to mitigate that risk? In this case, a biosafety level 3 laboratory alone will not control that risk. A BSL3 laboratory would mitigate the risk of release of a pathogen from the laboratory containment area, and provide some protection for workers inside that area. But a BSL3 laboratory does not, ipso facto, prevent cross-contamination. Instead, policies and procedures, and administrative controls (such as training and performance metrics) would play major roles in mitigating that particular risk. A comprehensive risk assessment would have both identified this risk and recognized that the BSL3's physical and engineering controls would have to be augmented by specific procedural and administrative controls.

The official CDC report explains that the work should have taken approximately three hours to complete. Because of time pressures to attend a meeting, and the heavy workload of the laboratory, the work in this case was done in half of that time. The CDC report indicates that a "standard protocol" was used to inoculate the H9N2 and H5N1 viruses into separate Madin–Darby canine kidney (MDCK) cell culture. The CDC scientist further described following a "best practices" protocol for temporal separation of low pathogenic avian influenza (LPAI) and high pathogenic avian influenza (HPAI) virus propagation. However, according to the CDC report, "this laboratory did not have a written, approved laboratory team-specific SOP for the work." Moreover, the scientist maintained no written documentation of his or her work inside the laboratory, "including the order in which the viruses were inoculated, which reagents were used, or how the BSC was decontaminated" (US Centers for Disease Control and Prevention 2014b).

Again, as Chapter 5 explains, practices and procedures—drawn directly from the risk assessment—should be designed and documented as an explicit element of the risk mitigation process. If the risk of cross-contamination were considered significant, then having a written, approved SOP for the work would be a high priority.

A particularly embarrassing aspect of this incident is that the mistake was first identified not by the CDC itself, but by the USDA laboratory that had received the contaminated H9N2 virus from the CDC. This reflects the absence of a performance metric system for this CDC laboratory, as described in Chapter 8. If the risk assessment had identified cross-contamination as a risk, the laboratory should have established performance indicators that would have regularly tested its operations against this specifically defined risk. In this case, so-called quality control measures should have been in place to test materials to exclude the presence of other organisms before transferring them outside of the laboratory. Notably, among its recommendations, the CDC report calls for "institute[ing] comprehensive quality control measures across all CDC laboratories." A biorisk management system would embed these performance metrics in the risk assessment and mitigation process, rather than impose the same requirement on all laboratories regardless of the risks.

Another recommendation in the CDC report is to "ensure that all Influenza Division staff are appropriately trained to understand when biosafety events are reportable and to whom" (US Centers for Disease Control and Prevention 2014b). The CDC scientists involved believed that a report to the CDC select agent responsible official was not necessary because no release had occurred; the H5N1 virus had been handled only in BSL3 select agent-registered facilities. However, the H5N1-contaminated H9N2 preparation had been transferred to CDC laboratories that were not select agent registered, and had been shipped outside of the CDC as if it were not a select agent. Clearly, better training on the regulatory requirements is necessary.

Although the CDC report emphasizes the need to improve its training on regulatory requirements, compliance training has limited effectiveness. As Chapter 6 explains, training should be built into the biorisk management system of a laboratory as another specific risk mitigation technique. The laboratory scientists and technicians should be trained on what the results of the risk assessment are, what physical and engineered controls are in place to address those risks, and what risks need to be mitigated primarily through procedural and administrative controls. In this case, training should have focused on the development of procedure-specific SOPs, record-keeping expectations, and performance metrics, as well as regulatory requirements. Training that is crafted specifically to reduce identified, unique risks will inevitably be more valuable than training on regulatory compliance.

CASE STUDY 3: MISHANDLING INVENTORY AT THE NIH

In early July 2014, a US Food and Drug Administration (FDA) scientist prepared a FDA laboratory in Building 29A on the NIH campus in Bethesda, Maryland for the lab's move to the FDA's new headquarters in Silver Spring, Maryland. During this work, the scientist found 12 cardboard boxes with 327 vials of various biological materials. The vials contained many dangerous agents, including those that cause dengue, influenza, Q fever, ricksettsia, and other unknown viruses. The material in 32 of the vials was immediately destroyed because 28 vials were labeled as normal tissue and 4 were labeled as "vaccinia." Six of the vials were labeled "variola," and 10 of the vials had unclear labeling but were suspected of containing smallpox (US Food and Drug Administration 2014).

The glass, heat-sealed vials were stored in a box with cotton padding. There was no evidence that any of the vials had been breached or that workers in the lab had been exposed to the materials. The vials appeared to date back to the 1946–1964 period. The smallpox vials were labeled with a specific date: February 10, 1954 (Dennis and Sun 2014a).

The 16 suspected smallpox vials were immediately secured in a containment laboratory on the NIH campus. The Federal Bureau of Investigation worked with the CDC and NIH to ensure safe packaging and secure transport of the 16 vials, which were flown to the CDC in Atlanta. Testing confirmed the presence of smallpox virus

DNA in 6 of the 16 vials. The smallpox was viable in at least two of the vials. The remaining 279 biological samples were transferred to the US Department of Homeland Security's National Bioforensic Analysis Center for safeguarding (US Food and Drug Administration 2014).

The CDC immediately notified the WHO, which has exclusive responsibility for the smallpox materials that are allowed by international law to reside in only two facilities, the CDC in Atlanta and the State Research Center of Virology and Biotechnology (VECTOR) in Novosibirsk, Russia. This was the first time that Variola major has been discovered outside these two facilities since the deadly virus was eradicated in 1979 (Sun and Dennis 2014).

In early September 2014, the NIH announced that it had discovered five additional improperly stored vials of select agents over the previous month. Three select agents were found at the NIH Clinical Center Department of Laboratory Medicine, including *Burkholderia pseudomallei*, *Francisella tularensis*, and *Yersinia pestis*, and ricin and Botulinum neurotoxin were found in other NIH laboratories. None of the agents was in a select agent laboratory or stored in conditions consistent with the select agent regulations (Collins 2014).

Also in early September 2014, the FDA announced that in July it had found vials of staphylococcal enterotoxin at a nonselect agent lab within the agency's Center for Food Safety and Applied Nutrition. The vials contained 8 milligrams of the toxin, three more than the quantity needed to be treated as a select agent. In both cases, the NIH and FDA reported the findings to the CDC select agent program, and relocated the materials into laboratories registered to store them (Dennis and Sun 2014b).

A ROLE FOR BIORISK MANAGEMENT?

Although the major media have characterized these discoveries of improperly stored select agents as safety incidents, they actually represent serious lapses in security. It is clear that the agents did not result in accidental exposures or releases that could have caused someone at the NIH to contract an infectious disease. However, the agents could have been stolen by someone with malicious intent, without the NIH's or FDA's knowledge, and subsequently used as a weapon to cause deliberate harm. The traditional biosafety system based on biosafety levels subordinates security issues, and often relegates security only to the implementation of the select agent regulations. Characterizing these events as safety incidents neglects the full spectrum of risks involved. By contrast, a biorisk management system considers all the risks to the bioscience facilities—risks from accidents, risks from intentional misuse, and risks from natural events—and prioritizes them against each other.

Yet the NIH has continued to describe these improperly stored pathogens only in the context of safety. In July, FDA employees did not receive an official communication about the discovery for seven days. One scientist, who worked in the building and spoke on the condition of anonymity for fear of retaliation, said he learned about it when his supervisor read a media report. An NIH spokeswoman in July said that officials did not notify employees about the discovery because the vials were checked and found to have no breaches (Sun and Dennis 2014). Yet, storing these agents in areas that lacked adequate security represented a potentially serious security vulnerability.

In September, Alfred Johnson, director of the NIH's Office of Research Services, which coordinated the NIH's "clean sweep" of its labs, told the *Washington Post*, "All of these [materials] were found in containers that were intact, and there have been no exposures" (Dennis and Sun 2014b).

Francis Collins, director of the NIH, sent a memorandum to his NIH colleagues in early September, acknowledging the discovery of the additional improperly stored select agents. Again, he focused only on the safety aspects of the situation, not the security risks or implications: "I want to reiterate that there were no personnel exposures associated with the storage or discovery of these vials or samples. There is no evidence to suggest that there was a safety risk to anyone in the lab, the surrounding area, or the community" (Collins 2014).

In late August, the NIH issued a "guide notice" in response to the "lapses in safety practices at federal laboratories" that indicated that NIH would initiate a National Biosafety Stewardship Month in September 2014, including a 24-hour stand-down to "reexamine current policies and procedures ... conduct inventories of infectious agents in all labs ... [and] reinforce biosafety training." The NIH said that the National Biosafety Stewardship Month would provide the opportunity for "scientists all around the country to reinforce existing practices; revisit existing guidelines and resources; optimize further biosafety oversight; and strengthen partnerships to achieve our shared biosafety goals" (National Institutes of Health 2014). In addition to the exclusive focus on safety, with seemingly no consideration for security, the tone of this guide notice reflects that existing NIH practices, guidelines, resources, and oversight are appropriate; they simply need to be reinforced, revisited, and optimized.

The FDA's lack of concern about these incidents is also troubling. In early July, at the time of the discovery of smallpox in the FDA laboratory, the FDA released an official statement on the incident that concluded: "Overlooking such a sample collection is clearly unacceptable" (US Food and Drug Administration 2014). Ironically, in mid-July, the FDA discovered staphylococcal enterotoxin in a laboratory that was not registered to store the material. But the FDA's acting chief scientist and FDA Director Margaret Hamburg were not informed of the finding of staphylococcus until almost three weeks later (Dennis and Sun 2014b).

As this book emphasizes throughout, a biorisk management system is based on the risk assessment concept and process. Risk assessments, by definition, ask what could go wrong. A comprehensive risk assessment at the NIH or the FDA would logically conclude that the following scenario would be significant risk: a dangerous select agent may exist in an area of the facility that is not adequately secured from theft or misuse, and the material could be found, stolen, and used maliciously. In this scenario, the NIH would be the source of a potentially serious or even devastating bioterrorism incident. The mitigation measures for such a scenario are fairly straightforward: ensure that no such "orphaned" select agents exist outside of limited access, secured areas. It seems that such a risk assessment was not conducted by the NIH or the FDA prior to August 2014, or the risk was not deemed high enough to warrant the implementation of the requisite mitigation measures.

CONCLUSION

The biosafety and biosecurity incidents that occurred at the CDC and the NIH in 2014 reveal that comprehensive biorisk management systems are likely not in place at these two premier US bioscience institutions. Not surprisingly, these facilities continue to rely on the current biosafety levels paradigm and the current US biosecurity regulations. However, these incidents demonstrate that the time has arrived for the US biosafety enterprise to consider alternative methods and systems.

What emerges from these brief case studies is the consistent absence of comprehensive biorisk assessments—documented studies of a specific activity, involving specific agents and a specific laboratory, conducting specific work, using specific equipment and protocols, and carried out by specific people. In a true biorisk management system, these comprehensive biorisk assessments should be the basis for determining experiment- or activity-specific risk mitigation measures, such as unique training, policies, procedures, and experiment- or activity-specific performance metrics. Explicitly defined roles and responsibilities, from top management to laboratory workers, and organizational commitment to safety and security, monitoring, evaluation, and continual improvement would be seamlessly integrated into laboratory operations as well. Had these sorts of programs and systems been in place, the incidents described in this chapter may not have happened, or at least they could have been identified much sooner.

Instead, only a minority of today's biosafety community execute substantive risk assessments, and even fewer comprehensively document those assessments. The majority, especially when we consider the broad international biosafety community, adopt an approach that assumes an agent's risk will be the same regardless of how it is used in the laboratory. This leads to generic risk assessments of an agent's inherent biochemical characteristics—what is really only a hazard assessment. The exact same material safety data sheets are used as risk assessments by hundreds of laboratories conducting thousands of different experiments or activities. These generic hazard assessments—assessing the hazard that an agent poses out of the context of its use—are translated into generic biosafety levels that are assumed to encompass adequate mitigation measures. Yet the diversity of the work is now so immense that these generic hazard assessments and generic levels of risk mitigation no longer suffice.

Another tendency in today's biosafety community—in the United States as well as overseas—is to relegate responsibility for safety and security to a so-called biosafety officer who has little or no management authority. This officer polices the laboratories and cites them for noncompliance with established rules and regulations. But ultimately, it is the principal investigator or laboratory manager who decides what safety and security measures to implement. One lab may choose to invest significant resources in safety and security; another laboratory in the same facility may not. The biosafety officer is delegated all the responsibility, but lacks any of the required authority or budget to promote individual or organizational behavior change. In a biorisk management system, by contrast, the entire culture of the organization embraces safety and security, and everyone's performance hinges on the effectiveness of the risk management system.

Often, the response to safety and security incidents such as these is to invest more resources in sophisticated engineering controls, or to require this work to be conducted only in even higher levels of containment. Either approach significantly increases the cost of science, and also implies that only the most highly resourced facilities can safely conduct work with dangerous agents. However, these case studies show that very well-funded laboratories that rely on the current biosafety paradigm make mistakes, and that additional engineering controls or higher levels of containment would not have made any difference. By contrast, a well-functioning biorisk management system is scalable to any bioscience facility in the world, from small hospital laboratories to large national institutes. Safety is not wholly dependent on the available resources or wealth of a facility, but on the facility's broad management commitment to safety.

Finally, the official blame for these incidents seems to be directed at individual employees or the specific laboratories involved. In many ways, the official government reports on these incidents are typical root cause analyses, aimed at identifying a single, initiating event. What is lost from such an incident event chain, as MIT scholar Nancy Leveson has explained, is that the environment that made that incident possible had been laid years before. If these specific events had not happened, would another similar incident have occurred sometime in the future (Leveson 2011)?

What has not yet emerged, but hopefully will eventually, is a recognition that there may be something wrong with the organizational culture* or management system that allowed these incidents to take place. Is there something amiss with the safety and security culture† in US bioscience facilities that makes these sorts of incidents possible? Could the implementation of a comprehensive biorisk management system, where the entire workforce, including executive management, is actively engaged in identifying and reducing biorisks, significantly improve the safety and security of bioscience facilities and operations in the United States and internationally?

REFERENCES

Collins, F.S. 2014. Interim Update on Comprehensive Sweep at the NIH. Memo to NIH staff. September 5. Published in Kaiser, J. NIH Finds More Forgotten Risky Pathogens and Toxins. *Science*, September 8, 2014. http://www.sciencenow.org/biology/2014/09/nih-finds-more-forgotten-risky-pathogens-and-toxins.

Dennis, B., and L.H. Sun. 2014a. FDA Found More than Smallpox Vials in Storage Room. *Washington Post*, July 16. http://www.washingtonpost.com/national/health-science/fda-found-more-than-smallpox-vials-in-storage-room/2014/07/16/850d4b12-0d22-11e4-8341-b8072b1e7348_story.html.

Dennis, B., and L.H. Sun. 2014b. More Deadly Pathogens, Toxins Found Improperly Stored in NIH and FDA Labs. *Washington Post*, September 5. http://www.washingtonpost.com/national/health-science/six-more-deadly-pathogens-found-improperly-stored-in-nih-and-fda-labs/2014/09/05/9ff8c3c2-3520-11e4-a723-fa3895a25d02_story.html.

---

* "Organizational culture" here refers to "the set of shared attitudes, values, goals, and practices that characterizes an institution or organization" (Merriam-Webster 2014).
† "Culture" here refers to "the set of values, conventions, or social practices associated with a particular field, activity, or societal characteristic" (Merriam-Webster 2014).

Leveson, N.G. 2011. *Engineering a Safer World: Systems Thinking Applied to Safety.*
     Cambridge, MA: MIT Press.
Merriam-Webster. 2014. http://www.merriam-webster.com/dictionary/culture.
Morgan, D. 2014. Exclusive: CDC Says Lab Director Behind Anthrax Mishap Resigns.
     Reuters News Service. July 23.
National Institutes of Health. 2014. Promoting Health, Science, and Public Trust through
     Laboratory Safety. Director's Blog. August 27. http://directorsblog.nih.gov/2014/08/27/
     promoting-health-science-and-public-trust-through-laboratory-safety/.
Russ, H., and J. Steenhuysen. 2014. CDC Reassigns Director of Lab Behind Anthrax Blunder.
     Reuters News Service. June 24.
Steenhuysen, J., and S. Begley. CDC Didn't Heed Own Lessons from 2004 Anthrax Scare.
     Reuters News Service. June 30.
Sun, L.H., and B. Dennis. 2014. Smallpox Vials, Decades Old, Found in Storage Room at
     NIH Campus in Bethesda. *Washington Post*, July 8. http://www.washingtonpost.
     com/national/health-science/smallpox-vials-found-in-storage-room-of-nih-campus-in-
     bethesda/2014/07/08/bfdc284a-06d2-11e4-8a6a-19355c7e870a_story.html.
US Centers for Disease Control and Prevention. 2014a. Report on the Potential Exposure to
     Anthrax. July 11.
US Centers for Disease Control and Prevention. 2014b. Report on the Inadvertent Cross-
     Contamination and Shipment of a Laboratory Specimen with Influenza Virus H5N1.
     August 15. http://www.cdc.gov/about/pdf/lab-safety/investigationcdch5n1contaminatio
     neventaugust15.pdf.
US Department of Health and Human Services. 2009. *Biosafety in Microbiological and
     Biomedical Laboratories.* 5th ed.
US Food and Drug Administration. 2014. Update on Findings in the FDA Cold Storage
     Area on the NIH Campus. July 16. http://www.fda.gov/newsevents/newsroom/
     pressannouncements/ucm405434.htm.

11 Future Development of Biorisk Management

Challenges and Opportunities

Benjamin Brodsky and Uwe Müeller-Doblies

ABSTRACT

As a discipline of its own, biorisk management has evolved rapidly over the past 10 years. Implementation of a systematic biorisk management approach remains uneven globally, and new, poorly understood biorisks continue to emerge that challenge traditional approaches to biosafety and biosecurity.* In this chapter, we discuss some of the challenges facing the discipline of biorisk management, and consider approaches to addressing them. Many of these challenges relate to the inherently diverse and variable nature of biological risks, the different existing systems to assess and address these risks, variability in risk tolerance at the local and international levels, and the availability of data and determination of criteria to support the selection and application of these systems within organizations that are operating in different geographical, social, and legal environments. This chapter will consider how the regional differences in assessing and controlling biorisks can be understood in a way that will eventually permit contextual comparison of biorisk management cornerstones based on the assessment, mitigation, and performance model.

INTRODUCTION

Over the past 10 years, laboratory biorisk management has evolved as a discipline in its own right, comprising biosafety, biocontainment, and laboratory biosecurity. Based on the recognition that prescriptive control systems for biological risks are too inflexible to effectively manage the rapidly evolving and changing biorisks that are emerging in an age of exponential biological discovery, regulators and facility operators (sometimes referred to as duty holders in a legal context) are exploring risk-based and performance-based approaches, including those that have been

---

* The term *biosecurity* is still widely used in two different settings. The first is the control and exclusion of alien species (both animal and plant), ranging from microbes to higher organisms. This is also referred to as (veterinary) biosecurity, or phytosanitary controls. More recently, (laboratory) biosecurity has been used to describe the controls applied to prevent the malicious misappropriation and release of biological agents.

successfully deployed in other high hazard industries. Following the development of the CEN Workshop Agreement 15793 in 2008, a number of international organizations have adopted or advocated for this type of approach to address biorisk management. Most recently, in May 2014, the Office Internationale des Epizooties (World Organisation for Animal Health) initiated replacement of the established classification of containment levels that has been in place for over 20 years with a new risk-based approach to laboratory biosafety and biosecurity (*vide infra*). This chapter examines the current *assessment-mitigation-performance* model for biorisk management and where opportunities for improvement may be found.

As the preceding chapters have illustrated, a risk-based approach informed by a careful and comprehensive *biorisk assessment* can deliver risk mitigation measures tailored to the particular risks at hand, thereby rendering them more risk commensurate and sustainable, and thus of better value. This notion is predicated on the assumption that the data is available to quantify the alternative risk paths leading to hazard release, and that the performance characteristics of alternative risk control systems are also available. However, despite many years of research, there are still significant gaps in our knowledge on some of the key properties of hazardous biological materials, and in the parameters of the risk control systems applied to mitigate the associated risks. Where this data is available, it provides evidence for a risk-based approach and supports a more cost-effective risk control system, but where data is lacking it can potentially stop work or make the work too costly. In many cases, the same risk control systems have been applied around the world without documented evidence that demonstrates their performance in the specific, local contexts in which they are being applied. This is especially true in the case of laboratory biosecurity, where detailed studies of the effectiveness of biosecurity control measures in various contexts are currently lacking. With a strong foundation in hazard identification and risk assessment, the biorisk management approach essentially compels facilities attentive to biosafety and biosecurity to give due consideration to the nature of the hazard, the nature of the work ongoing, and the multitude of factors—physical, human, environmental, legal, etc.—that may influence the likelihood and consequences of identified biosafety and biosecurity risks (biorisks) without relying on generic, prescriptive measures developed that do not consider the local risk context.

Risk mitigation (also referred to as *risk control*) based on assessments of institution-specific risks stands in contrast to prescriptive approaches, which offer facilities guidance on what to do but not necessarily why to do it. The biorisk mitigation measures implemented on the basis of sound biorisk assessments are more likely to address specific identified biorisks, improve operational efficiency, and importantly, improve understanding within the organization of the need for proper risk mitigation implementation. The cost to achieve these benefits is the investment needed to attain a much more thorough understanding of the biological hazards, the potential risk paths, consequences, and risk mitigation measures. It also requires regulators and organizations to determine accepted residual risk levels (the level of risk that is acceptable following the adoption of risk mitigation measures). Prescriptive approaches assume accepted risk as well, but often lack formal recognition or critical analysis of the acceptance of risk. These are often difficult to negotiate between all stakeholders in a society. Even in the countries with the most advanced regulatory requirements for biological risk control systems, the process of collective learning on how to build, operate, and maintain facilities in order to achieve and sustain the target risk profile is not easily achieved and is resource-intensive. The higher the risk concerned, the more resources that have to be allocated in order to deliver safe and secure risk mitigation systems, and to demonstrate their performance.

Finally, *performance* plays a key role in the biorisk management approach (see Chapter 8). Performance assessment, an often overlooked aspect of biorisk management in many facilities, is critical for ensuring that the biorisk management system is performing as intended, and for identifying opportunities for improvement. The development of new methods to ascertain performance, and compare the performance of various approaches to biorisk management, is key for further advancing the field.

This chapter will consider these issues and highlight challenges that require the input of the biorisk management community to effectively overcome, and for which effective solutions have either not been identified or not been globally implemented. Some of these challenges, such as financial resource constraints and infrastructure limitations, are often beyond the control of a given facility—but new, innovative approaches to biorisk management may help address these external limitations. In presenting a discussion of these challenges and opportunities, it is hoped that this chapter will contribute to positive action to address the more difficult aspects of biorisk management for which solutions have proven elusive to date.

CHALLENGES AND OPPORTUNITIES

The ongoing development and adoption of laboratory biorisk management by institutions around the world is encouraging. Looking ahead, the biorisk management community must identify and confront the major challenges that impede the further implementation of risk-based biosafety and biosecurity. Many of these challenges will require a significant investment, both financially and intellectually, to solve. The challenges and opportunities presented in this chapter are not a comprehensive account; rather, this is an attempt to identify a subset of critical challenges that face the community today, which may be considered rate limiting in the further advancement of the field. Identifying and broadly implementing solutions to these challenges could contribute significantly to the further development of the field, and to the establishment of safer and more secure bioscience facilities.

BIORISK ASSESSMENT

As discussed in Chapter 3, biorisk assessment is the bedrock of biorisk management. At its core, biorisk assessment is the process of identifying potential biohazards, threats, and associated biosafety and biosecurity risks, characterizing those risks in terms of likelihood and severity of the consequences, and determining whether the characterized risks are acceptable or not. In theory, the process is straightforward. In practice, however, risk assessment as instituted by the biorisk management community is a complex process with widely varying methodologies applied in the field, both formal and informal. Setting aside the general principles recounted in this book, no universally accepted approach to performing risk assessment exists—including the specific risk assessment process to be taken, who should perform the risk assessment, when and how frequently to perform risk assessments, whether and how to document, communicate, and review risk assessments, and what constitutes acceptable risk.

This diversity poses a significant challenge to the credibility of assessments and makes comparison between different risk assessments and risk mitigation systems difficult—perhaps even impossible. The criticism that biosafety and biosecurity guidance—for example, the guidance issued by the World Health Organization (WHO) and World Organisation for Animal Health (OIE)—is not being implemented globally is now being replaced with the more fundamental challenge that the biosafety and biosecurity risk assessments conducted by different institutions are not readily comparable.

Established resources in the field offer general guidelines on how to conduct risk assessments. However, Wagener et al. (2008) observed, "Although risk assessments are currently performed, the lack of a unified approach and appropriate tools makes such assessments unnecessarily difficult." Among the wide variety of risk characterization methodologies available, each approach has advantages and disadvantages. Among these, a number of formal risk assessment approaches have been developed and utilized. Organizations and individuals must determine which of these approaches, or combinations of approaches, to apply to their specific circumstances and the risks identified for assessment. Different practitioners may favor some of these approaches over others. Many organizations, particularly smaller organizations, do not employ any formal, documented risk assessment methodology at all, but instead rely on a combination of subject matter expert judgment, consultation, experience, common sense, available biosafety and biosecurity guidance, and basic pathogen information to make critical risk mitigation decisions. Thus, the cost of risk assessments, in terms of financial investment and the time required to complete them, can vary greatly.

This lack of comparability between risk characterization methods is a limitation of an approach to biosafety and biosecurity purely based on local risk assessment. Without the ability to compare their respective risk assessments, interfacility cooperation, including the exchange of staff, samples, and reagents between different institutions, can be quite a challenge.

The application of a wide variety of available risk characterization approaches is a result of several causes. First, as noted above, there is an obvious lack of a commonly accepted standard procedure, or set of procedures, for conducting risk assessment in the field. The diverse approaches to assessing biological risk also indicate that none of the approaches are fully satisfactory. The more formal, semiquantitative, and quantitative approaches can yield more consistent, reproducible, and comparable results, but are resource-intensive and require reliable data to realize their potential.

Risk analysts working in this field are also confronted with the challenging, but not unique, problem of having to consider ill-defined or wholly uncharacterized biological hazards. Unfortunately, biorisk management professionals often must confront new, naturally occurring hazards: recent notable examples include naturally occurring emerging pathogens, such as novel strains of influenza virus (H7N9, H1N1), which have sparked outbreaks in humans, and newly discovered viruses

(Middle East respiratory syndrome coronavirus (MERS-CoV), severe acute respiratory syndrome-associated coronavirus (SARS-CoV)). At least at the time of the initial emergence of new pathogens such as these, there is a paucity of critical biological data on which a biorisk assessment normally depends. From a biorisk management perspective, much remains unknown for many biological materials of concern, which further complicates the biorisk assessment process and introduces significant uncertainty. The traditional approach to this uncertainty is a precautionary approach, which applies more risk mitigation controls in the absence of evidence to support reliance on lower levels of control.

In addition, as biotechnology advanced as a field, the prospect of modifying, or even creating, known or previously unknown microorganisms has also become a commonplace reality. Like work with naturally occurring pathogens, the biorisks associated with the manipulation of genetically engineered (or modified) microorganisms must be considered and addressed. The close alignment between risk management approaches utilized in the laboratory manipulation of naturally occurring and genetically modified microorganisms has been noted previously, although frameworks for national oversight of work with these two classes of microorganisms may be distinct (Kimman et al. 2008).

A subset of these laboratory activities may be considered dual-use research of concern (DURC).* Several recent examples of research published in the literature have attracted considerable attention, including an ongoing debate surrounding so-called gain of function experiments involving influenza viruses (Roos 2014). The publication of these experiments has generated a sometimes intensive dialogue between and among researchers, biorisk management experts, national authorities, and the general public regarding the safety of these experiments and the potential for the misuse of experimental procedures by others. Understanding the safety and security risks associated with DURC is no less important than characterizing the biorisks for work with naturally occurring pathogens and other genetically engineered microorganisms. In some cases, some of the controversy surrounding DURC experiments may be attributed to the difficulty of accurately determining the likelihood and consequences of the risks associated with the work, due to the unknown or untested properties of the engineered microorganisms involved.

Predicting the properties of genetically engineered microorganisms, and particularly how these organisms will interact with human and animal hosts, is often difficult, time-consuming, and costly. This uncertainty complicates the risk-benefit analysis of performing research and determining appropriate biosafety and biosecurity mitigation measures for handling both naturally occurring and engineered microorganisms and toxins. While steps have been taken in recent years to develop risk assessment approaches for analyzing DURC, many practitioners have concluded that current risk assessment procedures require further enhancement. For example,

---

* The US government defines *DURC* as "life sciences research that, based on current understanding, can be reasonably anticipated to provide knowledge, information, products, or technologies that could be directly misapplied to pose a significant threat with broad potential consequences to public health and safety, agricultural crops and other plants, animals, the environment, materiel, or national security" (US Office of Science and Technology Policy 2013).

a recent report describing the discussions at a Wilton Park conference in 2013 on dual-use biology stated:

> Much work needs to be done to identify appropriate risk assessment factors relevant to DURC, taking into account the wide range of possible security concerns. In the future, a broader approach to risk could assess physical safety; economic security costs; diplomatic security; social and political stability; fear and anger and risk of research leading to the diminishing trust in government. It should also look at probability and take into account possible actors' motives as well as intelligence on terrorist actors. Current DURC risk assessments have been largely 'risk-benefit' analyses, and there is a need for much more comprehensive and quantitative risk assessments that specifically evaluate what could go wrong with certain research. (Wilton Park 2013)

Even for well-characterized pathogens, such as foot-and-mouth disease virus (FMDV), additional data could be used to better inform the risk assessment process. For example, FMDV is well recognized to be transmitted over long distances by air, and the aerosol infectious dose for 50% (ID50) can vary from 1 to 10 particles in cattle to several thousand infectious particles in pigs. However, there are still significant gaps in understanding how aerosols are generated and how the transmissibility can be influenced, by humidity, temperature, and particle size. As a result, there is little data to quantify the performance of risk mitigation measures for this virus.

The lack of quantitative data presents a more fundamental issue with regards to the performance of various risk mitigation (control) systems for the mitigation of biological risks (Kimman et al. 2008). As discussed in the Introduction, other high hazard industries have faced similar challenges and, in response to this challenge, have assembled data on the performance of safety-critical components and have agreed to reference methodologies for risk assessments, e.g., hazard operability studies and layer of protection analysis (LOPA). If the shared data is available to support the risk assessment process, even the more complicated methodologies can be carried out in a reasonable time frame. If the data is not available, facility operators and regulators are presented with a difficult task. The lack of data would direct the risk assessment toward a precautionary approach, which would add significantly to the cost of risk mitigation—including facility design, construction, and operation. In particular for pathogens of public health concern, a situation can occur where the benefit from the laboratory activity receives a higher priority than an evidence-based risk mitigation system, which outstrips the public health budget for the disease.

Because it may be difficult to measure the effectiveness of risk mitigation measures in managing biorisk, past experience suggests regulatory authorities will, in the absence of accepted standard risk mitigation systems, require demonstrated performance in line with the risk assessment. A natural tension may therefore emerge between accepted standard risk mitigation systems and alternative risk mitigation strategies where demonstrated performance data on the alternatives is lacking, such as in the implementation of new approaches to risk mitigation in the design and construction of new facilities. An example of this issue was the use of positive pressure suits in containment level (CL) 4 facilities in the United Kingdom (UK). Under the United Kingdom's Control of Substances Hazardous to Health regulations (COSHH

2002), a hierarchy of controls has to be applied. Based on the interpretation that suits are personal protective equipment—the lowest tier in the hierarchy of controls—air-fed positive-pressure suits were initially rejected as a control in laboratories operating at CL4. For this reason, only Class III cabinet-style CL4 laboratories have been approved for operation in the UK, although the option of suit laboratories where the work cannot be conducted in a cabinet line or isolator had been introduced into UK guidance in 2006 (Advisory Committee on Dangerous Pathogens, Health and Safety Executive 2006). However, when organizations intend to work in suited laboratories, they assume the full burden of proof and have to demonstrate to regulatory authorities that these suited systems are equally safe. Recently, there have been developments of new primary containment devices and research into the performance of positive-pressure encapsulating suits (Steward and Lever 2012).

Equally, risk acceptance, or the acceptable residual risk levels following risk mitigation determined by institutions for work to proceed, differs from industry to industry and from culture to culture. Risk acceptance is affected by many factors, both internal and external to the organization, as described in Chapter 3. For example, external stakeholders such as the general public, regulatory authorities, partner organizations, and others may have greatly diverging perceptions of risk based on local levels of awareness, the frequency and content of communications between stakeholders, media reporting, cultural factors, prior local and societal experiences involving particular pathogens and diseases, and so on. It is of paramount importance that this societal acceptance of risk be better defined and communicated, as it is an essential prerequisite for a risk-based approach to be successful. Unless this acceptable risk level is suitably defined and captured in local or national legislative instruments, facility operators may not be able to demonstrate that they have discharged their duties appropriately, and thus could remain open to legal challenges.

The observed diversity in the approaches to risk characterization and risk acceptance in different regions presents the biorisk management community with a significant challenge: *Is it possible to determine objectively whether a given risk assessment has been performed adequately? Just as practitioners of biorisk management must determine whether a given biosafety or biosecurity risk is acceptable, the acceptability of the risk assessment itself (in terms of suitability of the chosen methodology, quality and relevance of data, and data interpretation and analysis) must also be determined. Can a minimal bar—a standard set of objective criteria—be established by which a biorisk assessment, whether it be qualitative or quantitative, formal or informal, be evaluated by internal and external stakeholders for adequacy and completeness?*

In considering this challenge, legislators, regulators, and others charged with identifying biorisk assessment requirements have to maneuver the delicate balance between achievable substantiation of the biorisk assessment and the societal benefit expected from the work associated with the managed biorisks. If the required burden of proof becomes too onerous for an organization to meet, important work to protect livelihoods will be prevented, or diverted to organizations in localities that are either more risk tolerant or less risk conscious. In other words, the restrictions imposed by a risk-averse approach to laboratory biosafety and biosecurity can have unintended consequences if laboratory biorisks are considered too narrowly, without consideration given to the impact of these risks and measures to control them on other, related risks. The balance of laboratory biorisks against the management of other societal risk must be considered. For example, when developing biorisk management measures against a defined laboratory biorisk for a public health laboratory, managers must consider the potential impact of these risk mitigation measures on the ability of the laboratory to execute its public health mission in the event of a disease outbreak, and how that impact could influence the risk posed by an outbreak.

The societal acceptable risk from the laboratory activity involving a hazardous pathogen will significantly depend on the natural occurrence of the disease in the region/country (prevalence), and the tangible benefits of carrying out the laboratory activities related to the etiologic agent (Figure 11.1). As societal risk tolerance increases, for example, in the case of a common, endemic pathogen, the biorisk mitigation necessary at the laboratory to control laboratory biorisk associated with the pathogen decreases.

Referencing the acceptable risk (or target risk level) for a biocontainment laboratory operation against the actual prevalence of the disease in the local or regional

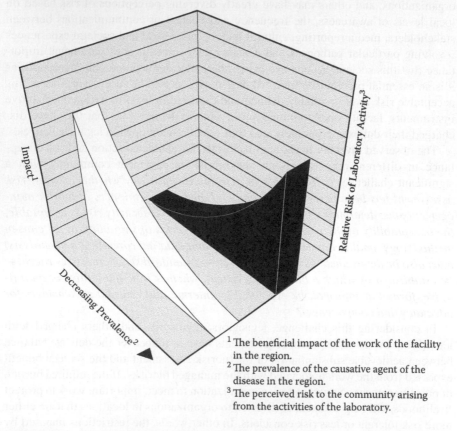

[1] The beneficial impact of the work of the facility in the region.
[2] The prevalence of the causative agent of the disease in the region.
[3] The perceived risk to the community arising from the activities of the laboratory.

FIGURE 11.1 Inverse relationship between societal risk acceptance and required biorisk mitigation.

settings enables realistic goal setting for biorisk management performance of laboratories while achieving an ethical balance between cost and benefit.

For example, the control measures taken in the laboratory to contain seasonal flu are less than those for historic flu strains not in circulation, even though the morbidity/mortality rate caused by each individual strain may be identical. The laboratory would have no impact on the prevalence of the seasonal flu strain, but is the single critical control point for the reintroduction of a historical flu strain (Wertheim 2010; Shoham 1993).

The biorisk management community should also consider whether certain defined risk assessment approaches, such as those mentioned above, are ideally suited for the assessment of certain categories or types of biosafety and biosecurity risks. This is particularly relevant for so-called low-probability (likelihood), high-consequence risks, which are often challenging to characterize in a quantitative fashion. Examples of such risks could include an accidental environmental release of a highly pathogenic agent, or the theft of a highly pathogenic agent for the purposes of terrorism.

If a standard risk characterization approach, or set of approaches, could be applied to a particular biosafety or biosecurity risk across all facilities globally, the expected output of these analyses will ideally be uniformly consistent across all facilities and regions, and thereby comparable. This is *not* to suggest that the risk likelihood and consequences, and associated risk acceptability, will be identical—but it will provide a stronger explanation for differences in these outputs.

Another great strength of conducting biorisk assessments specific to a particular setting as opposed to the approach of relying primarily on risk groups and corresponding biosafety level (BSL) categories is the ability to consider the consequences specific to the facility. For example, the endemicity of a particular disease/pathogen greatly influences the consequences of a pathogen release in a particular country or region—which will in turn affect risk acceptance. This can provide very sound reasoning for a different set of performance requirements for risk mitigation systems in different countries/regions. Given that a human life should have the same value wherever a laboratory is sited, some funders of facilities have insisted that they will only accept the risk control systems devised for a nonendemic zoonotic disease, even in countries where the disease is highly prevalent. At the same time, countries with relatively advanced laboratory infrastructure have not implemented such mitigation measures for pathogens like human immunodeficiency virus (HIV), even in times when there was no therapy available. A possible resolution to this conundrum is the relative risk approach. Rather than looking at the absolute risk posed by the facility itself, the risk relative to that of the naturally occurring disease is considered. If the disease is nonendemic, the risk of contracting the disease from the natural environment is very low, and thus the controls applied in the lab are relatively higher. If the disease is already prevalent in the country or region, the laboratory should ensure the risk arising from the laboratory activities, e.g., for laboratory workers and the public, is not greater than the risk in the community. This allows risk controls that are proportionate and more sustainable. For instance, it will mean that some countries in Africa may impose relatively higher risk controls on nonendemic viral diseases like Western equine encephalitis virus, but relatively lesser controls on Rift Valley fever virus or West Nile virus.

A similar approach has been taken with regards to laboratory biorisk management standards for foot-and-mouth disease in countries free of the disease. During a disease outbreak, the requirements for laboratory containment are reduced, but maintained at a level considered to ensure that the laboratory is not contributing to its spread. This reduces the burden on national reference laboratories that only handle the live virus in the event of a disease outbreak, but participate in proficiency testing schemes based on inactivated materials (European Commission for the Control of Foot-and-Mouth Disease 2013).

While the biorisk management community continues to develop its own risk assessment approaches, opportunities may exist for biorisk management professionals to interact with colleagues who work in other fields where novel materials are being generated and the risks of exposure of human and animal communities to these materials is unknown, such as nanomaterials toxicology. This interaction could help build, develop, and refine risk assessment methodologies that may better equip researchers to understand the most critical risks of proposed work with novel biological materials. The community should take proactive measures to encourage an exchange of views with risk analysts in other fields who have similar challenges and constraints in assessing risks associated with poorly defined hazards.

While risk analysts will continue to advance risk assessment methodologies, more must be done to ensure that all facilities that work with biological materials (naturally occurring or engineered) have tools and resources available to perform risk assessments. As indicated earlier, a one-size-fits-all approach to performing risk assessment is not feasible—different types of risks will warrant different approaches to analyzing them properly. A detailed assessment of the likelihood and consequences of a worker sustaining a needlestick injury and potential exposure to infectious material while executing a standard procedure in the laboratory will be performed much differently than the characterization of the risk of a large-scale release of biohazardous material into the environment due to a failure in the effluent waste treatment system, or the characterization of the risk of theft of sensitive data by an adversary determined to infiltrate a facility. However, at the facility level, it is necessary to compare these different risks in order to develop a balanced risk profile that permits the prioritization of resources for risk mitigation measures that address the highest relative risks.

A theme common to all biorisks is that they may be characterized by a risk path, which leads from a hazard to a hazard release, and subsequently to an adverse consequence. The key is empowering institutions to select appropriate risk characterization and communication tools to match the specific institutional needs. For example, bowtie risk control diagrams can complement risk assessments by systematically analyzing risk paths that lead to common hazard release events. They permit the comparison of similar types of risks within a facility to prioritize efforts for improvement. They have become a standard tool in the oil and gas industry and have also proven useful in biorisk management. A bowtie biorisk map can be used to help provide documentation to support management's decision-making process. See Figure 11.2 for an example of a biorisk bowtie model for the environmental release from a hypothetical laboratory working on exotic animal pathogens.

The risk paths can be divided into seven main categories with a number of individual risk paths, each depending on the laboratory activity and the laboratory

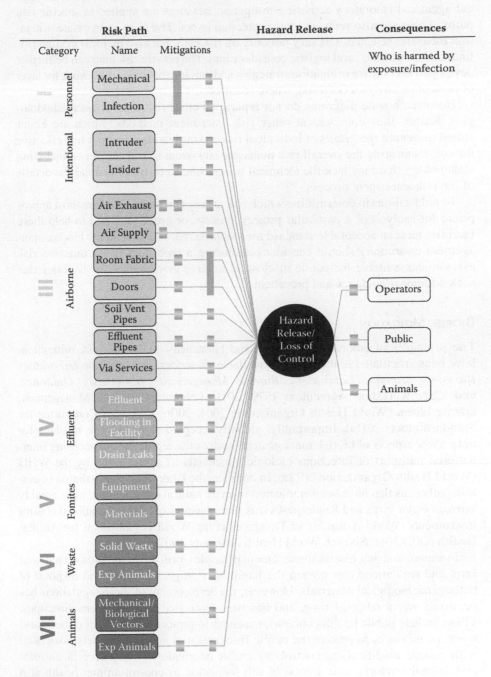

FIGURE 11.2 Basic bowtie diagram.

facility. Depending on the risk (likelihood and consequences) posed by the biological agent and laboratory activities, mitigation measures are applied to specific risk paths. These are also referred to as protection layers. The most appropriate mitigation measures selected will vary not only by risk path, but also by local availability, financial constraints, and logistic considerations. The bowtie risk map can be further developed to compare mitigation strategies and provide the logic framework for layer of protection analyses (LOPAs), where required.

However, bowtie diagrams do not replace all other risk assessment methodologies. Rather, they complement other risk assessment methods, which are better suited to capture the details of individual risk and mitigation steps, but less effective for communicating the overall risk management system to stakeholders—including stakeholders that may lack the technical background to fully understand the details of the risk assessment process.

To aid facilities in determining which risk assessment approaches are most appropriate for analysis of a particular project, process, or procedure, and to help these facilities meet an acceptable standard for a suitable risk assessment, the biorisk management community should consider establishing a repository of documented risk assessments, suitable for public disclosure, to serve as examples for facilities that seek additional guidance and precedent.

BIORISK MITIGATION

The principles of laboratory biosafety and biosecurity, and of biorisk mitigation, have been articulated in many freely available references, including the *Laboratory Biosafety Manual*, *Laboratory Biorisk Management: Biosecurity Guidance*, and CEN Workshop Agreement 15793:2011—Laboratory Biorisk Management, among others (World Health Organization 2004, 2006; European Committee for Standardization 2011a). Importantly, globally accepted standards are in place for only a few aspects of biorisk management, such as the regulations concerning international transport of infectious biological materials as summarized by the WHO (World Health Organization 2012a). In addition, the laboratory manipulation of certain pathogens that have been eradicated from the natural environment, most notably *Variola major* virus and Rinderpest virus, are subject to special international control instruments (World Assembly of Delegates of the World Organisation for Animal Health (OIE) 2011; Sixtieth World Health Assembly 2007).

In some countries, biorisk management principles form the foundation for national laws and regulations that govern the handling, storage, transport, and disposal of pathogenic biological materials. However, the development of these regulations has occurred over a range of time, and has been motivated from different directions. These include public health concerns, measures to protect the health of (laboratory) workers, measures to protect the public from potential negative impacts associated with genetic modification, controls to enable international commerce in animals and animal products, and animal health legislation to control animal health at a national level. As a result, there is currently a patchwork of national laws and regulations, with some countries having developed (over time) relatively robust national measures, while other countries may have little or no national measures in place to oversee work with biological materials. It is also difficult to directly compare different legal and regulatory structures across national boundaries. For example, while many countries utilize risk or hazard groups to classify biological materials (risk groups 1–4) and help determine appropriate risk mitigation measures, the exact definitions of each risk group vary between countries and between different competent authorities in the same country (American Biological Safety Association 2014a).

For these and many other reasons, at the global level, the current state of biorisk management implementation in biological facilities and during field operations is highly uneven. This is to be expected: not only are national legal requirements different, but the risks associated with different organizations and activities vary, risk perceptions among stakeholders also vary, and available resources to mitigate these risks are not equivalent. Some organizations struggle to translate the general principles of biorisk management into institution-specific policies, plans, and risk mitigation procedures, giving rise to a challenge: *despite significant progress, the biorisk management community faces a number of challenges in the journey toward establishing universal, risk-based standards and benchmarks for biorisk management systems that can accommodate the differences in biorisk and biorisk mitigation capacity at the local, national, and regional levels.*

The challenges associated with the disparate implementation of appropriate risk mitigation have long been recognized and cannot be solved overnight. Multiple standards exist that affect specific aspects of biorisk management, from laboratory design, to engineering controls, to personal protective equipment—too many to recount in detail here. Furthermore, the bioscience community continues to develop additional national and international standards, and methods to evaluate laboratory conformance to these standards, to harmonize biorisk management as a whole. A recent example is the development of ANSI/ASSE Z9.14-2014, "Testing and Performance Verification Methodologies for Ventilation Systems for Biosafety Level 3 (BSL-3) and Animal Biosafety Level 3 (ABSL-3) Facilities" (American Biological Safety Association 2014b). The benefits of increased standardization of biorisk management are numerous. They include a safer and more secure bioscience community through more effective biorisk management, facilitation of international scientific communication and research collaboration, and more effective infectious disease detection. Of course, efforts to establish universal standards and benchmarks are still relatively nascent, and significant investment is still required.

The need for an international performance-based standard in the field of biorisk management has been recognized by many. In 2008, CEN Workshop Agreement 15793:2008: Laboratory Biorisk Management (known as CWA 15793) was published. The document was the result of a workshop process that included 76 participants from 24 countries, as well as input from relevant international organizations. The document was renewed for a second three-year period in 2011 (European Committee for Standardization 2011a). In 2012, a related guidance document, CWA 16393:2012: Laboratory Biorisk Management—Guidelines for the Implementation of CWA 15793:2008 (CWA 16393), was published following the conclusion of another, similar workshop process (European Committee for Standardization 2012). CWA 16393 offers additional implementation guidance to organizations that seek to follow the requirements of CWA 15793.

CWA 15793 establishes 64 performance-based requirements for biorisk management systems. The requirements are intended to be compatible with any existing local or national regulations, as well as the requirements of other commonly used management system standards, such as ISO 9001:2008, ISO 14001:2004, and BS OHSAS 18001:2007. Together with CWA 16393, CWA 15793 gives organizations the performance-based criteria for establishing a comprehensive biorisk management system, which can be implemented using the assessment-mitigation-performance (AMP) model. In addition, CWA 15793 provides a useful set of definitions for terminology commonly utilized in the field of biorisk management. While initially intended for laboratories, the requirements of CWA 15793 are regarded by many as being suitable for any organization that works with biological materials or toxins, and those that seek to establish a formal and comprehensive biorisk management system. Importantly, like other management system standards, CWA 15793 does not establish specific prescriptive requirements; organizations are responsible for determining the manner in which the performance-based requirements of CWA 15793 will be fulfilled.

Since its publication, CWA 15793 has arguably become a standard reference for biorisk management professionals. In principle, the performance-based requirements may be applied by any organization, in any local setting, regardless of the specific biological hazards in use. The document does not take a risk management approach based on biological risk groups or laboratory biosafety levels, although the requirements of the document do not conflict with these more traditional approaches to laboratory biosafety and biosecurity. Rather, the document requires that organizations identify hazards and threats, and perform risk assessments. This design offers organizations the necessary flexibility to identify and implement specific biorisk management measures in a manner that satisfies the requirements, while also promoting greater effectiveness and sustainability than a purely prescriptive approach. Because the document is not prescriptive, organizations must carefully characterize and evaluate hazards and associated biorisks before implementing risk mitigation measures—an important precept of the AMP model for biorisk management. The document emphasizes key factors that an organization must put in place to support an effective biorisk management system. These include, for example, top management commitment to continual improvement of the system, the establishment of an organizational policy on biorisk management, the establishment of clearly defined roles and responsibilities, and the establishment of procedures for monitoring the biorisk management system, corrective action, and review. These factors, which are common to many well-known management system standards based on the plan-do-check-act (Deming) cycle (see more information in Chapters 2 and 8), are critical but sometimes overlooked elements of organizational biorisk management systems. Finally, the document includes a terms and definitions section, intended to help harmonize the usage of terminology in the field.

Despite being available for almost six years as of this writing, quantifiable data on the rate of adoption of CWA 15793 by organizations is lacking. A number of institutes are currently utilizing CWA 15793; for example, five institutions located in three European countries (France, the Netherlands, and Sweden) recently reported outcomes of a project aimed at implementing CWA 15793 within a new European

Laboratory Response Network (Sundqvist et al. 2013). However, the total number of organizations that have adopted the document, and how these organizations use the document, is unknown. No formal certification scheme exists based on CWA 15793; therefore, data concerning the number of organizational certifications against the requirements of the document is not available. The most comprehensive picture of CWA 15793 awareness and usage is provided by the European Biological Safety Association (EBSA). In 2013, EBSA administered a voluntary survey of its membership to examine the levels of awareness and implementation of CWA 15793 by its members.* The survey data suggests that there is a substantial level of awareness of the document within the European biorisk management community (85%), although a smaller proportion of the community reported that their organizations were currently adopting the requirements of the document (33%) (European Biosafety Association 2013). A majority of survey respondents expressed support for sustaining the document: 72% of respondents considered it as "important" or "very important" that CWA 15793 is maintained beyond 2014. This survey, which was focused on one region and one group (EBSA members), also highlights the need for a more thorough understanding of the global levels of awareness and adoption of the biorisk management systems approach described in CWA 15793.

Irrespective of the current adoption rate, it is clear that the development of CWA 15793 and the management systems approach to biorisk management has had a significant impact on contemporary laboratory biosafety and biosecurity. A prime example of this impact is the adoption by the World Assembly of Delegates of the Office Internationale des Epizooties (World Organisation for Animal Health) in May 2014 of a risk-based "Standard for Managing Biorisk in the Veterinary Laboratory and Animal Facilities" in the *OIE Terrestrial Manual 2014* (World Organisation for Animal Health 2014). This new standard is intended to support, and potentially replace, the OIE's previous biosafety and biosecurity guidance, which focused on the utilization of risk groups and containment levels. Instead, this new standard utilizes a risk-based approach aligned with CWA 15793. The World Health Organization has also made extensive use of CWA 15793—for example, through the publication in 2012 of a biosafety guidance manual for laboratories handling *Mycobacterium tuberculosis* (World Health Organization 2012b). The manual employs a risk-based approach that is based on the requirements of CWA 15793 and the technical guidance offered in the WHO *Laboratory Biosafety Manual*. Recently, CWA 15793 has also been suggested by the WHO as a key biorisk management resource for laboratories handling human specimens that may contain certain emerging pathogens, such as MERS-CoV (World Health Organization 2013).

Beginning in 2013, the International Organization for Standardization (ISO) began to consider adopting CWA 15793 to an ISO deliverable. The process now under way could ultimately lead to an ISO international standard for biorisk management. Many stakeholders in the biorisk management community are observing this

---

* In 2013, the American Biological Safety Association (ABSA) also administered a voluntary survey to its membership on CWA 15793. Preliminary results were presented at the 56th Annual Biological Safety Conference in October 2013; however, as of this writing the survey results have not been published.

process with interest. ISO produces several types of deliverables, each with unique requirements and characteristics (International Organization for Standardization 2014). It is possible that the conversion of CWA 15793 to an ISO deliverable could elevate the level of international awareness and adoption of performance-based biorisk management systems. The near-term future of formal standards in biorisk management along the lines of CWA 15793 and CWA 16393 will largely depend on the costs of adoption and benefits to biosafety and biosecurity that they provide to the bioscience community.

Should a voluntary international standard based on CWA 15793 emerge, another potential outcome is the eventual construction of an international certification scheme based on the standard. Already, at least one organization (the American Biological Safety Association (ABSA)) has established a relatively new voluntary laboratory accreditation scheme focused on laboratory biorisk management, and based in part on the requirements of CWA 15793 (American Biological Safety Association 2013). While the scope of this particular laboratory accreditation program is currently limited to high-containment laboratories located in the United States, it is conceivable that the lessons learned from the implementation of this accreditation program could inform the creation of a more global laboratory biorisk management certification scheme.

If properly implemented, a certification scheme based on an international biorisk management standard could afford organizations the opportunity to demonstrate an acceptable level of conformance with the standard based on a harmonized conformity assessment process, and thereby could potentially facilitate a greater degree of international collaboration between laboratories that work with biological materials. Should a certification scheme emerge, one area for future research will be to investigate the causal link between the achievement of biorisk management system certification based on an international biorisk management standard, and measurable improvements in actual biorisk management performance.

Regardless of what ultimately transpires via this process, the continued evolution of performance-based standards for biorisk management is a key issue for the biorisk management community. Additional information on the current adoption of CWA 15793 and CWA 16393, as well as analysis of what factors may inhibit the adoption of these documents by interested organizations, would be valuable contributions to the current dialogue.

In parallel with efforts to establish a biorisk management standard, efforts are underway to create universal benchmarks for education, training, and competency in biorisk management. For example, CEN Workshop Agreement 16335:2011: Biosafety Professional Competence, published in 2011, established professional competencies and training requirements for biorisk management professionals (European Committee for Standardization 2011b). To date, however, no internationally recognized system exists to assess and compare the competencies of biorisk management professionals who work in different laboratories. The International Federation of Biosafety Associations (IFBA) is now working to establish an international professional certification scheme to certify the competencies of biorisk management professionals, in part based on the stipulations of CWA 16335 (International Federation of Biosafety Associations 2012).

At the same time, more education and training opportunities in the field of biorisk management are becoming available. Several academic centers and biosafety associations now offer professional development opportunities for members of the biorisk management community. Despite this recent progress, these new professional development and certification opportunities are currently more readily accessible to professionals from relatively wealthy countries with greater opportunities to pursue university-level and professional-level education. The relative quality of educational and professional development opportunities in the biorisk management field is also difficult to assess and compare objectively. A challenge moving forward will be to build upon existing mechanisms for international communication and collaboration, such as those provided by national and international biosafety associations, to build links and professional networks between biorisk management professionals in countries with emerging biological laboratory infrastructure and their counterparts in more technically advanced countries. Innovative solutions are needed for building these partnerships, including those based on web-based education, training and mentoring, modern communication platforms such as social networking, joint training and research opportunities in the field, and other approaches. Cost-effective solutions that bridge geographic, language, and cultural barriers are clearly needed in order to build professional competency globally.

BIORISK MANAGEMENT PERFORMANCE—EVALUATING THE ORGANIZATIONAL BENEFITS FROM A MATURED BIORISK MANAGEMENT SYSTEM

If the implementation of a biorisk management system is not a regulatory requirement, institutions must be able not only to attain, but also to recognize tangible biosafety and biosecurity benefits in order to justify the investment of resources toward adopting the AMP model for biorisk management. CWA 15793 provides a framework that is increasingly used to measure biorisk management performance; a prominent example includes the use of CWA 15793 as the basis for the inspection protocol used by the WHO in 2009 and 2012 to assess the biosafety and biosecurity practices of the two laboratory repositories of Variola major virus (World Health Organization 2012c, 2012d). As the field continues to develop, compelling metrics for the impact of a biorisk management systems approach are needed to incentivize the commitment to the process. Organizations must recognize that the benefits of biorisk management may not be immediately obvious or readily measurable. Developing an organization from a risk management-naïve state to a risk management-intelligent one is a journey. It needs to be understood that the primary returns on investment in biorisk management are the accidents and incidents that have been prevented from happening, and are thus hard to count or quantify in monetary terms. However, the flexibility afforded to institutions by the AMP approach for designing tailored biorisk management systems may complicate efforts to compare and evaluate the sufficiency and performance of various organizational biorisk management systems, particularly if they are located in different countries or regions.

Biorisk management is a young conceptual approach, and little evidence has been published to date that proves the superiority of the AMP model for managing biorisk over more traditional and more established methods of biosafety and biosecurity. This is complicated by the fact that no mechanism exists to track biosafety- and biosecurity-related incidents and accidents at the global level, which, if available, could serve as a useful high-level barometer of the relative effectiveness of biorisk management systems based on the AMP approach. For example, publicly available data on laboratory-acquired infections (LAIs) is often scattered throughout the scientific literature, and probably underrepresents the true number of LAIs that have occurred globally, many of which may not be easily detected, characterized as LAIs, or reported publicly. While a number of important studies that examine the incidence of LAIs have been published in the past 20 years, the lack of comprehensive data renders difficult any attempt to compare or generalize incident rates across different laboratories, countries, and regions (Sewell 1995; Harding and Brandt Byers 2006; Walker and Campbell 1999). Relevant and reliable historic information regarding biosecurity incidents, such as thefts of biological materials, is even rarer. A recent study in the United States revealed that no incidents of theft of the most dangerous biological materials—so-called biological select agents and toxins*—from laboratories subject to stringent US biosecurity regulations had occurred between the years 2004 and 2010 (Henkel et al. 2012). While encouraging, this result applies only to a subset of US biological laboratories, and does not necessarily reflect a global trend toward stronger laboratory biosecurity. More data is needed here as well. In addition, as alluded to at the outset of this chapter, the consequences of increased laboratory biosecurity regulations—for example, the potential impacts on the beneficial exchange of reagents and biological agents for the development of diagnostic proficiency and interlaboratory comparison of diagnostic performance—need to be considered and weighed against the potential biosecurity risks.

The rates of occurrence of less serious incidents, such as near misses that do not result in significant adverse consequences, but offer opportunities to identify and address contributing factors that could result in accidents, are also not known. Subject to the thresholds used for reporting, these low-level occurrences could be a better metric for improvement. However, they may be underreported even internally within organizations because of embarrassment, fear of negative repercussions for the worker(s) involved, or local safety culture. Other biorisk management performance indicators, such as those discussed in Chapter 8, may serve as effective proxies for reliable incident data to support our understanding of biorisk management performance. However, these performance indicators are not yet standardized across the biorisk management field. This currently hampers the use of performance indicators to measure the impact of biorisk management system adoption relative to other risk control approaches.

The dearth of reliable incident-related data and resultant improbability of observing measurable declines in overall incident rates at the global level, as well as the

---

* The list of biological select agents and toxins is available at http://selectagents.gov/Select Agents and Toxins List.html (accessed October 1, 2014).

current lack of standardized, publicly available performance indicators for biorisk management, give rise to a final challenge: *novel means of providing a more complete understanding of the absolute and relative effectiveness of biorisk management system implementation in driving biorisk reduction at both the organizational and global levels are urgently needed.*

This challenge is not unique to the evaluation of the effectiveness of biorisk management. As discussed in the Introduction, performance-based risk management approaches have been adopted by organizations in several high hazard industries as a means to improve safety performance and business continuity. Prominent examples of safety management systems include those adopted by the chemical, nuclear energy, and offshore oil and gas drilling industries (Committee on the Effectiveness of Safety and Environmental Management Systems 2012). At the same time, management systems approaches for improving occupational health and safety of workers, such as the requirements embodied in the international occupational health and safety specification BS OHSAS 18001, have become more common. It is clear that the development of biorisk management has been strongly influenced by the widespread adoption of similar approaches in these other fields. We may attempt to predict the impact of biorisk management adoption by examining evidence of the impact these other, more established and more widely implemented management systems have had in their respective industries.

For example, several studies have attempted to gauge the impact of occupational health and safety systems (such as BS OHSAS 18001 adoption) on health and safety in the workplace. Interestingly, according to one literature review, despite several years of study, the effectiveness of occupational health and safety management systems still has not been rigorously proven (Robson et al. 2007). The authors of the review acknowledge, however, that when considered together, prior studies do offer indications that adoption of these management systems can have a positive impact on worker health and safety. A more recent study draws a similar conclusion, although the study's authors note that "additional research is required to assess possible causal relationships between SMS [safety management system] implementation and the improvement in safety level of a company" (Bottani et al. 2009). These examples offer a cautionary lesson: proving the impact of the adoption of biorisk management on biorisk control will likely require several years of careful study.

Of course, biorisk management is not only concerned with reducing occupational health risk to workers, but also focused on managing safety and security risks to the facility and the broader community. In this sense, the historic performance of other management systems designed to control safety and security risks, including those noted above, may offer lessons for biorisk management.

What opportunities exist for the biorisk management community to address this challenge? Certainly, as discussed in previous chapters, there is an urgent need to develop leading and lagging safety and security performance indicators for AMP-based biorisk management systems. To the extent possible, the development of these performance indicators should be coordinated and standardized across the community to facilitate comparative and reliable biorisk management performance measurement. The biorisk management community could leverage existing approaches to defining performance indicators now utilized in other sectors in developing and standardizing performance indicators. The UK Health and Safety Executive has led the development of a framework of safety performance indicators (SPIs) for containment level 4 laboratories (Atkins 2011) based on the methodologies for the establishment of performance indicators for process safety in the nuclear and chemical industry (Organization for Economic Cooperation and Development 2008). The development of performance indicators will facilitate understanding of how the AMP model for biorisk management contributes to safer and more secure operations. Should an international standard for biorisk management eventually be developed, performance indicators could also play an important role in internal and external conformity assessment processes (such as audits) based on the standard's requirements.

As part of this effort, the biorisk management community should consider how to better capture trends in biosafety and biosecurity incident rates over time. While the establishment of an international database or reporting mechanism for such incidents, such as LAIs, is likely infeasible in the near term, it may be possible to establish test beds to experiment with various mechanisms that could be used to capture this information. For example, the creation of a voluntary LAI reporting mechanism involving a group of participating laboratories may offer a means to measure and analyze LAI incident rate data more systematically than what is currently possible. The biorisk management community could explore approaches that have previously been employed in related sectors. For example, since 1991, the International Healthcare Worker Safety Center at the University of Virginia in the United States has administered a program known as the Exposure Prevention Information Network (EPINet). According to the EPINet website, participating healthcare organizations voluntarily share incidents where percutaneous injuries or contact with blood and body fluids occurred within their facilities (International Healthcare Worker Safety Center 2014). Similar, voluntary mechanisms could be employed for the reporting of a broader set of biosafety and biosecurity incidents.

Finally, the biorisk management community should undertake rigorous, scientific studies to collect and analyze evidence to support or refute the hypothesis that AMP-based biorisk management systems reduce biorisks to laboratory personnel, facilities, and communities in a more cost-effective and sustainable manner than more traditional, prescriptive, and compliance-driven approaches to biosafety and biosecurity. As studies conducted on occupational health and safety management systems demonstrate, this hypothesis—while entirely rational and plausible—is not necessarily simple to confirm scientifically. The biorisk management community must be careful to avoid conflating the plausibility of improved effectiveness of the AMP model with proof of improved effectiveness. The importance of conducting in-depth scientific studies to critically evaluate this hypothesis cannot be understated. Nothing less than the safety and security of laboratory workers, facility operations, and community health, and perhaps even national and international health security, stands to be impacted. With so much at stake for the bioscience community, additional evidence-based approaches to evaluate how AMP-based biorisk management systems impact the safety and security of laboratories, their workers, and the public are essential.

CONCLUSION

The term *biorisk management* was coined around 2007 as part of the development of the CEN Workshop Agreement on laboratory biorisk management—CWA 15793. The AMP approach on which biorisk management rests—including the concept of increasing the stringency of risk management measures to meet increased risk—appealed to most stakeholders, as concerns were raised from the outset that a formal management systems approach to biosafety and biosecurity would require a significant administrative effort that would not necessarily be warranted to the same degree for low-risk organizations. In the high hazard pathogen sector, there was a clear recognition that there was a lack of structure to the management controls applied in the sector, resulting in the technical standards not always implemented and adhered to as intended. In parallel, there was an uneven set of requirements with regards to laboratory biosecurity (physical security, inventory control, staff security, and information security) because of differing legal and regulatory requirements in different countries. The management systems approach promised to deliver management arrangements that could be audited externally and would enable the facility owners to demonstrate that the risks associated with the operation of a bioscience facility were recognized and managed effectively. This was and is a very important goal—working toward this goal ultimately builds the currency of trust among facility operators, laboratory workers, regulators and other authorities, and the wider public.

The future of the AMP-based model for biorisk management is therefore bright. The approach is increasingly perceived and accepted as an effective and appropriate means for bioscience organizations to assess and control biorisk, regardless of the hazards they handle or the resources available to them. The biorisk management community cannot become complacent, however. Despite broad praise for the approach, the total number of organizations who have successfully implemented a comprehensive biorisk management system in line with all requirements of CWA 15793 is still small, and does not appear to be growing rapidly. Developing risk management maturity across organizations in the industry is necessary for advancing the biorisk management systems approach. There is enough support to initiate a transition of the CWA 15793 into an ISO management system document, but as discussed here, there are a number of challenges that contribute to the slow implementation at the organizational level. In this chapter, we have outlined what we believe are a few of the most significant of these challenges. These included challenges related to biorisk assessment, biorisk management system implementation, and the evaluation of biorisk management system performance, particularly in comparison to other approaches to controlling biosafety and biosecurity risks.

An important issue that must be taken into account is how the biorisk management approach described in CWA 15793 sits in relation to the formal regulatory approach currently in place in many regions. The legal and regulatory framework in a given locality influences the capacity and tolerance that facility operators may have for additional administrative overhead related to laboratory biorisk management system implementation. Transition to a regulatory approach based on biorisk management will also require local authorities to determine they are sufficiently resourced to implement and oversee a risk-based biorisk management framework, as such a framework could require a higher level of resources and technical competency. And while the community most impacted by biorisk management is supportive of a risk-based approach for the benefits that have been discussed in the chapters of this book, there is still more stakeholder communication required to obtain broader support, particularly among the public as a stakeholder.

Addressing these challenges will minimize the likelihood of future events involving biological materials, including major incidents. To do so, the biorisk management community must continue to improve the tools for biorisk management practitioners. This includes better tools for risk assessment and communication, means to identify and demonstrate the tangible benefits that reward an organization for the implementation of the formal biorisk management systems approach (especially at the lower end of the hazard spectrum where biosafety and biosecurity risks may carry lesser consequences), and methods to better measure and compare the performance of biorisk management systems. As more organizations measure their biorisk management performance in relation to biorisks, the data will become available to generate the evidence base to improve organizational efficiency and streamline costs associated with risk mitigation, while further minimizing risk. This will enable the biorisk management community to continue creating tangible benefits for the bioscience community, including keeping society and the environment safe while more efficiently facilitating the delivery of science.

REFERENCES

Advisory Committee on Dangerous Pathogens, Health and Safety Executive. 2006. *Biological Agents: The Principles, Design and Operation of Containment Level 4 Facilities.* Health and Safety Executive (UK).

American Biological Safety Association. 2013. ABSA High Containment Laboratory Accreditation Program. July. http://www.absa.org/aiahclap.html (accessed May 1, 2014).

American Biological Safety Association. 2014a. Risk Group Classification for Infectious Agents. http://www.absa.org/riskgroups/index.html (accessed April 2, 2014).

American Biological Safety Association. 2014b. ANSI/ASSE Z9.14-2014 Standard: Background Materials. http://www.absa.org/resansiasse.html (accessed July 1, 2014).

Atkins, WS. 2011. *Development of Suitable Safety Performance Indicators for Level 4 Bio-Containment Facilities: Phase 2.* Health and Safety Executive (UK).

Bottani, E., L. Monica, and G. Vignali. 2009. Safety Management Systems: Performance Differences between Adopters and Non-Adopters. *Safety Science,* 47: 155–162.

Committee on the Effectiveness of Safety and Environmental Management Systems for Outer Continental Shelf Oil and Gas Operations, Transportation Research Board. 2012. *Transportation Research Board Special Report 309: Evaluating the Effectiveness of Offshore Safety and Environmental Management Systems.* Washington, DC: National Academies Press.

Control of Substances Hazardous to Health Regulations (COSHH). 2002. United Kingdom. http://www.legislation.gov.uk/uksi/2002/2677/contents/made.

European Biosafety Association. 2013. Survey on the Awareness and Usage of Laboratory Biorisk Management CWA 15793:2011 and Its Guidance Document CWA 16393:2012. Brussels.

European Commission for the Control of Foot-and-Mouth Disease. 2013. Minimum Biorisk Management Standards for Laboratories Working with Foot-and-Mouth Disease Virus. April 22–24. http://www.fao.org/fileadmin/user_upload/eufmd/40thGeneral_session_documents/40General_Session/App7_RFMD_Minimumstandard.pdf (accessed September 25, 2014).

European Committee for Standardization. 2011a. *CEN Workshop Agreement 15793:2011: Laboratory Biorisk Management*. Brussels.

European Committee for Standardization. 2011b. *CEN Workshop Agreement 16335:2011: Biosafety Professional Competence*. Brussels.

European Committee for Standardization. 2012. *CEN Workshop Agreement 16393:2012: Laboratory Biorisk Management—Guidelines for the Implementation of CWA 15793:2008*. Brussels.

Harding, A.L., and K. Brandt Byers. 2006. Epidemiology of Laboratory-Associated Infections. In *Biological Safety: Principles and Practices*. Washington, DC: ASM Press, p. 53.

Henkel, R.D., T. Miller, and R.S. Weyant. 2012. Monitoring Select Agent Theft, Loss and Release Reports in the United States—2004–2010. *Applied Biosafety*, 17(4): 171–180.

International Federation of Biosafety Associations. 2012. IFBA Certification Program.

International Healthcare Worker Safety Center, University of Virginia. 2014. EPINet. http://www.healthsystem.virginia.edu/pub/epinet/about_epinet.html (accessed January 9, 2014).

International Organization for Standardization (ISO). 2014. ISO Deliverables. http://www.iso.org/iso/home/standards_development/deliverables-all.htm (accessed April 2, 2014).

Kimman, T.G., E. Smit, and M.R. Klein. 2008. Evidence-Based Biosafety: A Review of the Principles and Effectiveness of Microbiological Containment Measures. *Clinical Microbiology Reviews*, 21(3): 403–425.

Organization for Economic Cooperation and Development. 2008. *Guidance on Developing Safety Performance Indicators Related to Chemical Accident Prevention, Preparedness and Response: Guidance for Industry*. 2nd ed. Paris.

Robson, L.S., et al. 2007. The Effectiveness of Occupational Health and Safety Management System Interventions: A Systematic Review. *Safety Science*, 45: 329–353.

Roos, R. 2014. Experts Call for Alternatives to 'Gain-of-Function' Flu Studies. May 22. http://www.cidrap.umn.edu/news-perspective/2014/05/experts-call-alternatives-gain-function-flu-studies (accessed July 3, 2014).

Sewell, D.L. 1995. Laboratory-Associated Infections and Biosafety. *Clinical Microbiology Reviews*, 8(3): 389–405.

Shoham, D. 1993. Biotic-Abiotic Mechanisms for Long Term Preservation and Re-Emergence of Influenza Type A Virus Genes. *Progress in Medical Virology*, 40: 178–192.

Sixtieth World Health Assembly. 2007. Smallpox Eradication: Destruction of *Variola* Virus Stocks. Presented at WHA 60.1, Geneva, May 18.

Steward, J.A., and M.S. Lever. 2012. Evaluation of the Operator Protection Factors Offered by Positive Pressure Air Suits against Airborne Microbiological Challenge. *Viruses*, 4: 1202–1211.

Sundqvist, B., U.A. Bengtsson, H.J. Wisselink, B.P. Peeters, B. van Rotterdam, E. Kampert, S. Bereczky, N.G. Johan Olsson, A. Szekely Björndal, S. Zini, S. Allix, and R. Knutsson. 2013. Harmonization of European Response Networks by Implementing CWA 15793: Use of a Gap Analysis and an "Insider" Exercise as Tools. *Biosecurity and Bioterrorism: Biodefense Strategy, Practice, and Science*, 11(Suppl. 1): S36–S44.

US Office of Science and Technology Policy. 2013. United States Government Policy for Institutional Oversight of Life Sciences Dual Use Research of Concern. February 22.

Wagener, S., A. Bennett, M. Ellis, M. Heisz, K. Holmes, J. Kanabrocki, J. Kozlovac, P. Olinger, N. Previsani, R. Salerno, and T. Taylor. 2008. Biological Risk Assessment in the Laboratory: Report of the Second Biorisk Management Workshop. *Applied Biosafety*, 13(3): 169–174.

Walker, D., and D. Campbell. 1999. A Survey of Infections in United Kingdom Laboratories, 1994–1995. *Journal of Clinical Pathology*, 52: 415–418.

Wertheim, J.O. 2010. The Re-Emergence of H1N1 Influenza Virus in 1977: A Cautionary Tale for Estimating Divergence Times Using Biologically Unrealistic Sampling Dates. *PLoS One*, 5(6): e11184.

Wilton Park. 2013. *Conference Report—Dual-Use Biology: How to Balance Open Science with Security.*

World Assembly of Delegates of the World Organisation for Animal Health (OIE). 2011. Resolution No. 18: Declaration of Global Eradication of Rinderpest and Implementation of Follow-Up Measures to Maintain World Freedom from Rinderpest (Adopted by World Assembly of Delegates of the OIE). May 25.

World Health Organization. 2004. *Laboratory Biosafety Manual.* 3rd ed. Geneva: World Health Organization Press.

World Health Organization. 2006. *Biorisk Management: Laboratory Biosecurity Guidance.* Geneva: World Health Organization Press.

World Health Organization. 2012a. *Guidance on Regulations for the Transport of Infectious Substances: 2013–2014.* Geneva: World Health Organization Press.

World Health Organization. 2012b. *Report of the World Health Organization (WHO) Biosafety Inspection Team of the Variola Virus Maximum Containment Laboratories to the Centers for Disease Control and Prevention (CDC): Atlanta, Georgia, USA, 7–11 May 2012.* Geneva: World Health Organization Press.

World Health Organization. 2012c. *Report of the World Health Organization (WHO) Biosafety Inspection Team of the Variola Virus Maximum Containment Laboratories to the State Research Centre of Virology and Biotechnology ("SRC VB VECTOR"), Federal Service for Surveillance on Consumer Rights.* Geneva: World Health Organization Press.

World Health Organization. 2012d. *Tuberculosis Laboratory Biosafety Manual.* Geneva: World Health Organization Press.

World Health Organization. 2013. *Laboratory Biorisk Management for Laboratories Handling Human Specimens Suspected or Confirmed to Contain Novel Coronavirus: Interim Recommendations.* Geneva: World Health Organization Press.

World Organisation for Animal Health (OIE). 2014. Standard for Managing Biorisk in the Veterinary Laboratory and Animal Facilities. In *OIE Terrestial Manual 2014.* May.

Index